D0998248

# Cannibalism

Also by Bill Schutt

*Dark Banquet: Blood and the Curious Lives of Blood-Feeding Creatures*

# Cannibalism

## A Perfectly Natural History

BILL SCHUTT

ALGONQUIN BOOKS OF CHAPEL HILL 2017

Published by
Algonquin Books of Chapel Hill
Post Office Box 2225
Chapel Hill, North Carolina 27515-2225

a division of
Workman Publishing
225 Varick Street
New York, New York 10014


Printed in the United States of America.
Published simultaneously in Canada by Thomas Allen & Son Limited.
Design by Steve Godwin.



Library of Congress Cataloging-in-Publication Data
Names: Schutt, Bill, author.
Title: Cannibalism : a perfectly natural history / by Bill Schutt.
Description: First edition. | Chapel Hill, North Carolina : Algonquin Books of Chapel Hill, 2017. | "Published simultaneously in Canada by Thomas Allen & Son Limited."
Identifiers: LCCN 2016023112 | ISBN 9781616204624
Subjects: LCSH: Cannibalism. | Cannibalism—Cross-cultural studies.
Classification: LCC GN409 .S38 2017 | DDC 394/.909—dc23
LC record available at https://lccn.loc.gov/2016023112

10 9 8 7 6 5 4 3 2 1
First Edition

For Janet and Billy Schutt

And for my best friend,
Robert A. Adamo (1953–2011)

# Contents

# Prologue

*A census taker tried to quantify me once. I ate his liver with some fava beans and a big Amarone.*

—Thomas Harris, *The Silence of the Lambs*

To mark its 100-year anniversary in 2003, the American Film Institute polled a jury of 1,500 actors, writers, directors, and historians, to determine the 50 greatest screen villains of all time. Topping the AFI list was the ultimate in fictionalized cannibals, Dr. Hannibal "The Cannibal" Lecter. In *The Silence of the Lambs*, Jonathan Demme's Academy Award–winning vision of the Thomas Harris novel, Lecter, memorably portrayed by Sir Anthony Hopkins, helps recently graduated FBI recruit Clarice Starling track down "Buffalo Bill," a serial killer who skins his female victims in order to tailor a "woman suit."

Second place in the poll went to Norman Bates, the mother-fixated hotel manager inhabiting Alfred Hitchcock's 1960 classic *Psycho*. Okay, I know what you're thinking: *Norman Bates wasn't a cannibal*, but just give me a minute.

From the opening scene, Hitchcock invited Eisenhower-era audiences to indulge in some long-held taboos. Filmgoers titillated the previous year by the first of the Rock Hudson/Doris Day bedroom comedies suddenly found themselves transformed into voyeurs, peering into shadowy corners previously unseen by mainstream movie audiences of the 1950s. From an amorous lunch-hour rendezvous (where the half-clad lovers had obviously just risen from their unmade hotel room bed) to a peephole in the Bates Motel, nobody would be confusing Hitchcock's masterpiece with *Pillow Talk*.

Released to a mixed critical response, the movie became a sensation with audiences, and remains so today. More than a half-century after its release, Bernard Herrmann's strings-only score is perhaps the most instantly recognizable music ever written for a film. Less well known is the fact that Joseph Stefano's screenplay for *Psycho* had been adapted from a Robert Bloch pulp novel about Wisconsin native Edward Gein (pronounced *Geen*), a real-life murderer, grave robber, necrophile, and cannibal.

Born in 1906, Gein lived a solitary and repressive life under the thumb of a domineering mother. The family owned a 160-acre farm, seven miles outside the town of Plainfield, but when his brother died in 1944, Gein abandoned all efforts to cultivate the land. Instead, he relied on government aid and the occasional odd job to support himself and his mother. When she died in 1945, Gein found himself alone in the large farmhouse, sealing off much of it and leaving his mother's room exactly as it looked when she was alive. The house itself fell into such serious disrepair that the neighborhood kids began claiming that it was haunted.

On the night of November 17, 1957, things began to unravel for the recluse known as Weird Old Eddie. The police were investigating the disappearance of local storeowner Bernice Worden when they got a tip that Gein had been seen in her hardware store several times that week. They picked him up at a neighbor's house where he was having dinner and questioned him about the missing woman. "She isn't missing," Gein told them, "she's down at the house now."

Gein's dilapidated farmhouse had no electricity, so the cops used flashlights and oil lamps to pick their way through the debris-laden rooms. In a shed out back, one of the men bumped into what he thought were the remains of a dressed-out deer hanging from a wooden beam. But the fresh carcass hanging upside down was no deer: It was the decapitated body of Mrs. Worden. As the stunned lawmen moved through the gruesome crime scene, it became clear that the neighborhood kids had been right. The Gein house was haunted. Each room they entered presented them with a new nightmare: soup bowls made from human skulls, a pair of lips attached to a window shade drawstring, and a belt made from human nipples. In the kitchen, the police reportedly found Bernice Worden's heart sitting in a frying pan on the stove and an icebox stocked with human organs.

Soon after Gein's arrest, media correspondents from all over the world began descending on the town and its shocked populace. The reporters poked around the Gein farm and interviewed neighbors. Some of the locals recounted how they'd been given "venison" by Gein, who later told authorities that he had never shot a deer in his life. The Plainfield Butcher had also been a popular babysitter.

With the publication of a seven-page article in *Life* magazine (and a three-page spread in *Time*), millions of Americans became fascinated with Ed Gein and his crimes. Plainfield became a tourist attraction with bumper-to-bumper traffic crawling through the narrow streets. Jokes called "Geinisms" became popular.

Q: What did Ed Gein give his girlfriend for Valentine's Day?

A: A box of farmer fannies.[1]

The following year, Robert Bloch loosely adapted the Gein crimes for his novel, relocating his tale to Phoenix and concentrating on the mother-fixation aspects of the story while playing down the mutilation and cannibalism. An assistant gave Alfred Hitchcock the book and he procured the film rights soon after reading it. The director also had his staff buy up as many copies of the novel as they could find. He wanted to prevent readers from learning about the plot and then revealing its secrets. After some initial resistance from Paramount Pictures, the "Master of Suspense" directed his most famous and financially successful film—one that would never have been made if not for Ed Gein, a quiet little cannibal, who explained to the authorities, "I had a compulsion to do it."[2]

Is it really a surprise, though, that our greatest cinematic villain is a man-eating psychiatrist while the mild-mannered runner-up is based on a real-life cannibal killer? Perhaps not, if one considers that many cultures share the belief that consuming another human is the worst (or close to the worst) behavior that a person

---

1 In the 1940s and 1950s Fanny Farmer was the largest producer of candy in the U.S.

2 When *Psycho* opened on June 16, 1960, it was an instant hit, with long lines outside theaters and broken box office records all over the world. More than 50 years later the film is remembered best for its famous shower scene, one which reportedly caused many of our Greatest Generation to develop some degree of ablutophobia, the fear of bathing (from the Latin *abluere*, "to wash off"). Few theatergoers realized that the "blood" in *Psycho* was actually the popular chocolate syrup, Bosco (a fact the company somehow neglected to mention in their ads and TV commercials).

can undertake. As a result, real-life cannibalistic psychopaths like Jeffrey Dahmer (another Wisconsin native) and his Russian counterpart, Andrei Chikatilo, have attained something akin to mythical status in the annals of history's most notorious murderers. Whether through a filter of fictionalization, where man-eating deviants are transformed into powerful antiheroes, or through tabloids sensationalizing the crimes of real-life cannibals, these tales feed our obsession with all things gruesome—an obsession that is now an acceptable facet of our society.

A different attitude was taken toward "primitive" social or ethnic groups whose members might not have shared the Western take on cannibalism taboos. At best, these "savages" were pegged as souls to be saved, but only if they met certain requirements. In the first half of the 20th century, for example, explorers and the missionaries who followed them ventured into the foreboding New Guinea highlands and quickly imposed one hard-and-fast rule for the locals: Cannibalism in any form was strictly forbidden.

But far worse instances of cultural intrusion occurred elsewhere and throughout history, as those accused of consuming other humans, for any reason, were brutalized, enslaved, and murdered. The most infamous example of this practice began during the last years of the 15th century when millions of indigenous people living in the Caribbean and Mexico were summarily reclassified as cannibals for reasons that had little to do with people-eating. Instead, it paved the way for them to be robbed, beaten, conquered, and slain, all at the whim of their new Spanish masters.

Similar atrocities were carried out on a massive scale by a succession of flag-planting European powers who (if one believes their accounts) wrested South America, Africa, and the Indo-Pacific away from man-eating savages, whose behavior placed them beyond the pale of anything that could remotely be described as human.

So were European fears about cannibalism simply an invention used to justify conquest, or were there cultures, including those encountered by the Spaniards, where the consumption of humans was regarded as normal behavior? Although defining someone as a cannibal became an effective way to dehumanize them, there is also evidence that ritual cannibalism, as embodied in various customs related to funerary rites and warfare, occurred throughout history.

As I began studying these forms of cannibalism, I sought to determine not only their perceived functions, but just how widespread they were or weren't. Surprisingly—or perhaps not surprisingly given the subject matter—there is disagreement among anthropologists regarding ritual cannibalism. Some deny that it ever occurred, while others claim that the behavior did occur but was uncommon. Still others claim that cannibalism was practiced by many cultures throughout history and for a variety of reasons. One such body of evidence led me straight back to European history, where I learned that a particularly macabre form of cannibalism had been practiced for hundreds of years by nobility, physicians, and commoners alike, even into the 20th century.

As a zoologist, I was, of course, intrigued at the prospect of documenting cases of non-human cannibalism. Looking back now, I can see that I'd started my inquiry with something less than a completely open mind. Part of me reasoned that since cannibalism was presumably a rare occurrence in humans (at least in modern times), it would likely be similarly rare in the animal kingdom.

Once I dug further, though, I discovered that cannibalism differs in frequency between major animal groups—nonexistent in some and common in others. It varies from species to species and even within the same species, depending on local environmental conditions. Cannibalism also serves a variety of functions, depending on the cannibal. There are even examples in which an individual being cannibalized receives a benefit.

In several instances, cannibalism appears to have arisen only recently in a species, and human activity might be the cause. In one such case, news reports informed horrified audiences that some of the most highly recognizable animals on the planet were suddenly consuming their own young. "Polar bears resort to cannibalism as Arctic ice shrinks," reported CNN, while the *Times of London* echoed the sentiment: "Climate Change Forcing Polar Bears to Become Cannibals." It was Reuters, though, that scored a perfect ten on the gruesome scale with an online slide show in which an adult polar bear was seen carrying around the still cute-as-a button head of a dead cub, the remains of its spinal cord trailing behind like a red streamer.

The real story behind polar bear cannibalism turned out to be just as fascinating, though it would also serve as a perfect example of how many accusations of and stories about cannibalism throughout history were untrue, unproven, or exaggerated—distorted by sensationalism, deception, a lack of scientific knowledge, and just plain bad writing. With the passage of time, these accounts too often become part of the historical record, their errors long forgotten. Part of my job would be to expose those errors.

I was also extremely curious to see if the origin of cannibalism taboos could be traced back to the natural world, so I developed a pair of alternative hypotheses. Perhaps our aversion to consuming our own kind is hardwired into our brains and as such is a part of our genetic blueprint—a gene or two whose expression selects against such behavior. I reasoned that if such a built-in deterrent exists, then humans and most non-humans (with the exception of a few well-known anomalies such as black widow spiders and praying mantises) would avoid cannibalism at all costs. Thus, the taboo would have a biological foundation.

Conversely, I weighed the possibility that the revulsion most people have at the very mention of cannibalism might stem solely from our culture. Of course, this led to even more questions. What are the cultural roots of the cannibalism taboo and how has it become so widespread? I also wondered why, as disgusted as we are at the very thought of cannibalism, we're so utterly fascinated by it? Might cannibalism have been more common in our ancestors, before societal rules turned it into something abhorrent? I looked for evidence in the fossil record and elsewhere.

Finally, I considered what it would take to break down the biological or cultural constraints that prevent us from eating each other on a regular basis. Could there come a time, in our not-so-distant future, when human cannibalism becomes commonplace? And for that matter, was it already becoming a more frequent occurrence? The answers to these questions are far from certain but, then again, there is much about the topic of cannibalism that cannot be neatly divided into black and white. Likely or not, though, the circumstances that might lead to outbreaks of widespread cannibalism in the 21st century are grounded in science, not science fiction.

My aim was to stay away from the clichéd ideas about cannibalism that are already ingrained in our collective psyche and, with such a wealth of relevant material to explore, I quickly realized that this wouldn't be difficult. Even the most famous cannibal stories, it turned out, had factual gaps that are only now being filled. In the case of the Donner Party, for example, I joined researchers whose scientific approach to the most infamous cannibalism-related event in American history had shed new light on this 19th-century tale of stranded pioneers.

I've tried to approach each example from a scientific viewpoint, delving into what I considered the most intriguing aspects of anthropology, evolution, and biology to provide the broadest yet most engaging natural history of this behavior. What happens to our bodies and minds under starvation conditions? Why are women better equipped to survive starvation than men? And what physiological extremes would compel someone to consume the body of a friend or even a family member?

With regard to criminal cannibalism (Jeffrey Dahmer and his ilk), I was less interested in the overhashed and gory details of the crimes than the reasons for our enthrallment with the overhashed and gory details. This is not a book that explores the minds of our so-called cannibal killers, though it does seem that instances of cannibalism-related crime may be on the uptick. I've also taken a hard line on sensationalism by highlighting and differentiating between physical evidence, ethno-history, unfounded information, and horse feathers.

In the pages ahead, you will encounter everything from cannibalism *in utero* to placenta-munching mothers who carry on a remarkably rich tradition of medicinal cannibalism. Yes, the ick factor is high, but I hope you'll find this journey as fascinating and unusual as I have—a journey whose goal is to allow us to better understand the complexity of our natural world and the long and often blood-spattered history of our species.

With this in mind, why not grab a glass of red wine, and let's get started.[3]

---

3 For suitable background music, for starters I suggest "Timothy", the catchy one-hit-wonder by The Buoys. The song, written by Rupert Holmes ("The Piña Colada Song"), tells the tale of three trapped miners, two of whom survive by eating the title character. In 1971 "Timothy" reached number 17 on the Billboard Top 100, even though many major radio stations refused to play it. In an unsuccessful attempt to reverse the ban, executives at Scepter Records began circulating a rumor that Timothy was actually a mule.

# Cannibalism

# 1: Animal the Cannibal

*Cannibals prefer those who have no spines.*

—Polish science fiction writer Stanislaw Lem, *Holiday*, 1963

I was knee-deep in a temporary pond that seemed to be composed of equal parts rainwater and cow shit when the cannibals began nibbling on my leg hair.

"If you stand still for long enough, they'll definitely nip you," came a voice from the shore.

The "they" were cannibalistic spadefoot toad larvae (commonly known as tadpoles) and the warning had come from Dr. David Pfennig, a biology professor at the University of North Carolina who had been studying these toads in Arizona's Chiricahua Mountains for more than 20 years.

At Pfennig's invitation, I had arrived at the American Museum of Natural History's Southwestern Research Station in mid-July—just after the early-summer monsoons had turned cattle wallows into nursery ponds and newly hatched tadpoles into cannibals. But the real reason I had come to the ancestral land of Cochise and the Chiricahua Apaches wasn't because the tadpoles were eating each other. It was because some of them *weren't* eating each other. In fact, when this particular brood had hatched about a week earlier, they were all omnivores, feeding on plankton and the suspended organic matter referred to in higher-class journals as "detritus."

Then, two or three days later, something peculiar took place. Some of the tiny amphibians experienced dramatic growth spurts, their bodies ballooning in size overnight. Now, as I waded, scoop-net in hand, through Sky Ranch Pond (a slimy-bottomed mud hole with delusions of grandeur), the pumped-up proto-toads were four or five times larger than their poop-nibbling brethren.

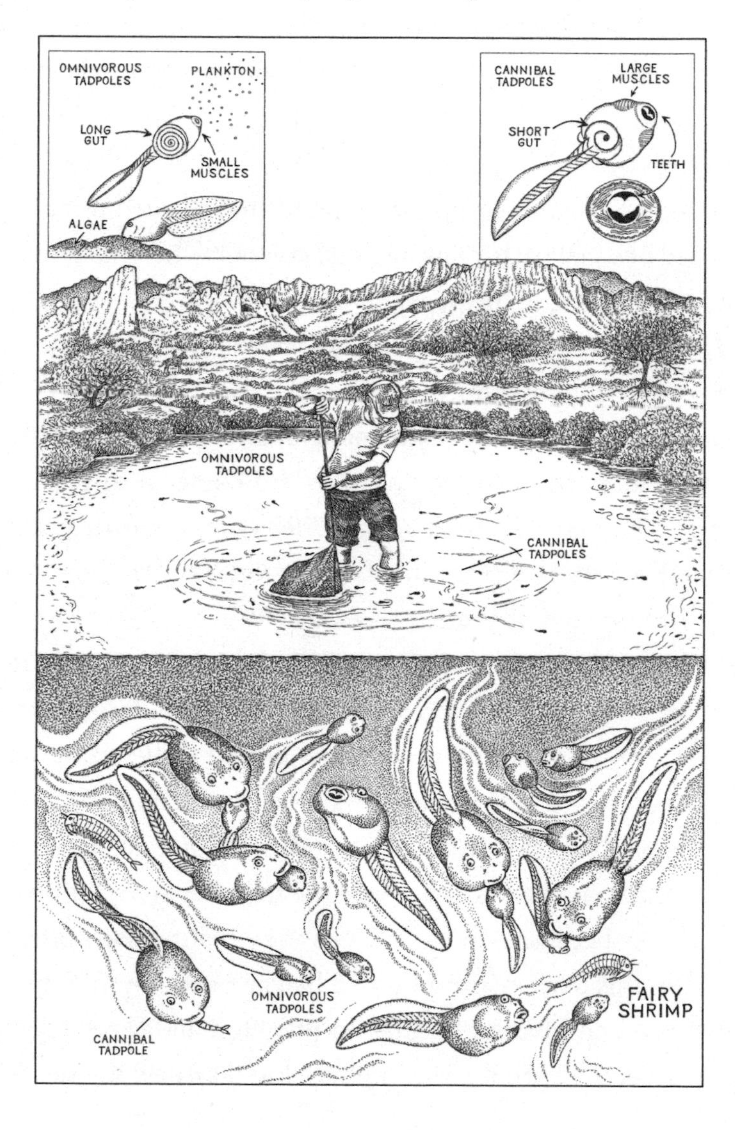

"These look like two different species," I said, examining a handful of tadpoles that I'd just scooped up. I also noted that the larger individuals were light tan in color while the little guys had bodies flecked with dark green.

"Initially, people thought they were different species," Pfennig replied.

Using a magnifying glass to get a better look at my squirmy captives, I saw that the differences went beyond body size and color. The larger tadpoles were also sporting powerful tails and serious-looking beaks.

"Yikes, nice choppers," I commented, always the scientist.

"They're made of keratin," Pfennig said. This was the same tough, structural protein found in our nails and hair.

Later, while comparing the two tadpole morphs under a dissecting 'scope, I saw that behind a set of frilly lips, the flat keratinous plates (which worked fine for detritus dining) had been transformed into a jack-o-lantern row of sharp-edged teeth in the cannibalistic forms. It was also evident that the jaw muscles were significantly enlarged in the cannibals, especially the jaw-closing *levator mandibulae*, whose bulging appearance reminded me of a kid with six pieces of Dubble Bubble jammed into each cheek (a dangerous behavior I only rarely attempt anymore ). Studies had shown that myofibers, the cells making up these muscles, were larger and greater in number (or hypertrophied and hyperplasious, respectively)—producing a more powerful bite. Of course, the extra bite force was necessary because, beyond latching onto the occasional unshaved human leg, these critters were using bulked-up bodies and the weaponry that accompanied it to subdue and consume their omnivorous pondmates.

Not quite so obvious was a significant shortening of the gastrointestinal (GI) tract in the cannibals, with the explanation relating to the dietary differences that accompanied the tadpole transformations. In the omnivores, a long GI tract is required for the breakdown of tough-to-digest plant matter, while a shorter GI tract works just fine when the diet is a fleshy one.[4]

Over a three-day period, I watched and captured tadpoles in bodies of water that ranged from tire-carved puddles to bovine swimmin' holes of the double-wide Olympic variety. Accompanied at various times by Pfennig, his wife, Karin, their two young daughters, and a pair of extremely personable UNC grad students, Antonio Serrato and Nick Levis, I learned a great deal about the three species of *Spea* that laid their eggs in such dangerously unpredictable conditions. Much of this information centered on the ecology, behavior, and evolution of these creatures. Of course, the cannibalism angle was there as well, although these researchers (including the kids) treated that particular behavior as perfectly normal.

Until relatively recently though, and with a very few exceptions, cannibalism in nature would have been regarded as anything but normal. As a result, until the last two decades of the 20th century,

---

4 A lengthy intestine is a hallmark of many herbivores, since longer guts translate to longer passage times for the food moving through them—allowing for additional chemical digestion and more absorption of nutrients. In many animals, though (including all vertebrates and even inverts like termites), the digestive tract cannot digest cellulose, the polysaccharide that makes up plant cell walls. The problem is solved by the presence of endosymbiotic bacteria or protozoans ("gut flora") that produce cellulases—enzymes capable of digesting polysaccharides. In "foregut fermenters" like cows, a multi-chambered stomach serves as a homestead for the enzyme-generating microbial horde, while in "hindgut fermenters" like horses, a pouchlike section of the intestine called the cecum houses the endosymbionts.

few scientists spent time studying a topic thought to have little, if any, biological significance. Basically, the party line was that cannibalism, when it did occur, was either the result of starvation or the stresses related to captive conditions.

It was as simple as that.

Or so we thought.

IN THE 1970S, Laurel Fox, a University of California Santa Cruz ecologist, took some of the first steps towards a scientific approach to cannibalism. She had been studying the feeding behavior of predatory freshwater insects called backswimmers (belonging to the order Hemiptera, the "true bugs"). Fox determined that, while the voracious hunters relied primarily on aquatic prey, "cannibalism was also a consistent part of their diets."

I contacted Fox and asked her about the transition that had taken place in the scientific community regarding this behavior. She told me that her observations in the field had sparked her interest and that, soon after, she began compiling a list of scientific papers in which cannibalism had been reported. Although there turned out to be hundreds of references documenting the behavior in various species, no one had linked these instances together or come up with any generalizations regarding the behavior. By the time Fox's review paper came out in 1975, she had concluded that cannibalism was not abnormal behavior at all, but a completely normal response to a variety of environmental factors.

Significantly, Fox also determined that cannibalism was a far more widespread occurrence than anyone had previously imagined and that it took place in every major animal group, including many that were long considered to be herbivores . . . like butterflies. She emphasized that cannibalism in nature, which some researchers referred to as "intraspecific predation," also demonstrated a complexity that seemed to match its frequency. Fox suggested that the occurrence of cannibalism in a particular species wasn't simply a "does occur" or "doesn't occur" proposition, but was often dependent on variables like population density and changes in local environmental conditions. Fox even followed cannibalism's environmental connection onto the human branch of the evolutionary tree. After pondering reports that humans practicing non-ritual cannibalism lived in "nutritionally marginal areas," she proposed that consuming other humans might have provided low-density populations with 5 to 10 percent of their protein requirements. Conversely, she suggested that cannibalism was rare in settlements where populations were dense enough to allow for the production of an adequate and predictable food supply.

In 1980, ecologist and scorpion expert Gary Polis picked up the animal cannibalism banner and began looking at invertebrates that consumed their own kind. Like Fox, he noted that while starvation could lead to increases in the behavior, it was certainly not a requirement. Perhaps Polis's most important contribution to the subject of cannibalism in nature was assembling a list of cannibalism-related generalizations under which most examples of invertebrate cannibalism could be placed. 1) Immature animals get eaten more often than adults; 2) Many animals, particularly invertebrates, do not recognize individuals of their own kind, especially eggs and immature stages, which are simply regarded as a food source; 3) Females are more often cannibalistic than males; 4) Cannibalism increases with hunger and a concurrent decrease in alternative forms of nutrition; and 5) Cannibalism is often directly related to the degree of overcrowding in a given population.

Polis emphasized that these generalizations were sometimes found in combination, such as overcrowding and a lack of alternative forms of nutrition (a common cannibal-related cause and effect), both of which now fall under the broader banner of "stressful environmental conditions."[5]

In 1992, zoologists Mark Elgar and Bernard Crespi edited a scholarly book on the ecology and evolution of cannibalism across diverse animal taxa. In it, they refined the scientific definition of *cannibalism* in nature as "the killing and consumption of either all or part of an individual that is of the same species." Initially the researchers excluded instances where the individuals being consumed were already dead or survived the encounter—the former they considered to be a type of scavenging. Eventually, though, they decided these were variants of cannibalistic behavior observed across the entire animal kingdom. Although there are certainly gray areas (encompassing things like breastfeeding or eating one's own fingernails), my fallback definition of *cannibalism* for this book is: The act of one individual of a species consuming all or part of another individual of the same species. In the animal kingdom, this would include behavior like scavenging (as long as the scavenged body was from the same species as the scavenger) and maternal care in which tissue (i.e., skin or uterine lining) was consumed. In humans,

---

5 Tragically, Dr. Polis drowned when his research vessel sank during a storm in the Sea of Cortez in 2000, an accident that also claimed the lives of a graduate student and three Japanese ecologists.

cannibalism would extend beyond the concept of nutrition into the realms of ritual behavior, medicine, and mental disorder.

As the study of cannibalism gained scientific validity in the 1980s, more and more researchers began looking at the phenomenon, bringing with them expertise in a variety of fields. From ecologists we learned that cannibalism was often an important part of predation and foraging, while social scientists studied its connection to courtship, mating, and even parental care. Anatomists found strange, cannibalism-related structures to examine (like the keratinous beak of the spadefoot toad), and field biologists studied cannibalism under natural conditions, thus countering the previous mantra that the behavior was captivity-dependent.

By the 1990s, Polis's generalizations had been observed among widely divergent animal groups, both with and without backbones, supporting the conclusion that the benefits of consuming your own kind could outweigh the often substantial costs. Once these generalizations became established, and as a new generation of researchers built upon foundations constructed by pioneers like Fox and Polis, cannibalism in nature, with all of its intricacies and variation, began to make perfect evolutionary sense.

Arizona's lowland scrub stood in stark contrast to the lush peaks and bolder-strewn valleys of the state's Chiricahua Mountains. These "sky islands" (isolated mountains surrounded by radically different lowland environments) provided a spectacular backdrop for my afternoon wade through yet another transient pond.

The air temperature had risen to 95 degrees Fahrenheit, which kept most of the area's terrestrial denizens hiding in shade or below ground, but the inhabitants of Horseshoe Pond reminded me of sugared-up kindergarteners tearing around a playground (albeit with fewer legs and more cannibalism). By this time, I had already begun to see distinct patterns of behavior in the spadefoot tadpoles that motored hyperactively just below the water's surface.

I noticed that the smaller, omnivorous morphs generally stuck to the shallows bordering the shoreline. They buzzed through the brown water in a non-stop, seemingly random quest for food, changing direction abruptly and often. One explanation for the patternless swimming behavior became apparent as I waded farther away from the shore, for here in the deeper water was the realm of the cannibals. I stood quietly and watched as hundreds of conspicuously larger tadpoles crisscrossed the pond, making frequent excursions from the deeper water toward the shore in a relentless search for prey.

*Immature animals get eaten more than adults*, I thought. Certainly, although in this case the youngsters were eating each other.

"They remind me of killer whales hunting for seals," said Ryan Martin, a former student of Pfennig's, now a professor at Case Western Reserve, who was also studying spadefoot toads here in Arizona.

Martin's comparison was spot on, and I threw him a nod, watching as a tiny hunter swam away from the shore with an even tinier pondmate clamped tightly between its serrated jaws.

So why did the local spadefoot larvae exhibit cannibalistic behavior? There certainly seemed to be enough organic matter suspended in these algae-tinted ponds to feed the entire brood and more.

As I spoke to Pfennig and his team of researchers, I learned that the answer was directly linked to the aquatic environments in which the adult amphibians deposited their egg masses.[6] Formed by spring and early-summer monsoons, the transient ponds frequented by the spadefoots (spadefeet?) are often little more than puddles, and as such they can evaporate quite suddenly in the hot, dry environment of southeastern Arizona. Natural selection, therefore, would favor any adaptations enabling the water-dependent tadpoles to "get out of the pool" as quickly as possible (i.e., to grow legs). In this instance, the phenomenon that evolved can be filed under the rather broad ecological heading of *phenotypic plasticity*: When changing environmental conditions allow multiple phenotypes (observable characteristics or traits) to arise from a single genotype (the genetic makeup of an organism). To clarify this concept, here are a couple of non-cannibalism-related examples.

Water fleas (Daphnia) are tiny aquatic crustaceans, named for a swimming style in which they appear to jump like fleas. In response to the appearance of backswimmers (Laurel Fox's favorite predatory insects), Daphnia develop tail spikes and protective crests. Although the genetic potential for body armor is always there (in the Daphnia's genotype), it doesn't exhibit itself until a specific environmental change occurs, in this case the arrival of Daphnia-munching backswimmers.

Here's another example, unrelated to cannibalism. The reef-inhabiting bluehead wrasse (*Thalassoma bifasciatum*) is famous for its habit of removing parasites from much larger fish, even entering

6 Toads and frogs belong to the amphibian order Anura (from the Greek for "no tail"). Most anurans lay their eggs in fresh water, with hatchlings undergoing complete metamorphosis from gill-bearing tadpoles to lung-breathing juveniles.

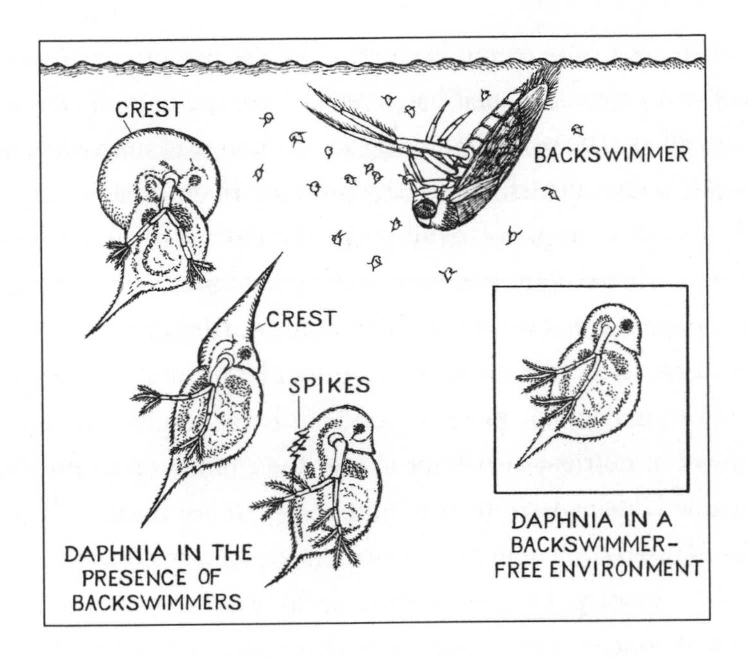


into their mouths. In this case, however, it's the removal of a male wrasse from its harem of 30 to 50 females that alters their local environment. Rather than waiting for a new male to arrive, something extraordinary takes place in the harem. Within minutes, one of the females begins exhibiting male-typical behaviors. Relatively quickly, the former female transforms into a male, a form of phenotypic plasticity known in the trade as *protogyny*. The opposite occurs in *protandry*, in which individuals begin life as males and transform into females. Examples include the clownfish (*Amphiprion*), whose behavior could have offered an intriguing alternative resolution to the animated film *Finding Nemo*.

In spadefoot toads, though, it's not the appearance of a predator or the loss of a harem's personal sperm bank that initiates the alternate phenotype (i.e., cannibalistic larvae). The selection pressure lies in the temporary nature of the brood ponds, where the eggs are deposited and hatch and where the tadpoles develop into toadlets. The period from egg to juvenile toad normally takes around 30 days unless, that is, the pond dries out first, killing the entire brood. In response to this particular environmental selection pressure, what evolved was a means by which some of the tadpoles can mature in about two-thirds of the time (20 days). The increased growth rate occurs because the cannibal larvae are getting a diet high in animal protein as well as a side order of veggies, the latter in the form of nutrient-rich plant matter their omnivorous prey had consumed during what turned out to be their last meal.

In an interesting note, *Spea couchii* does *not* transform into cannibalistic morphs but has evolved an alternative solution to the transient pond problem. Couch's spadefoot can go from egg to toad in only eight days—an amphibian record.

Though the story of spadefoot toad cannibalism has been well researched, it is not fully resolved. The reason is that no one has been able to identify the precise stimulus within these brood ponds that triggers the appearance of the cannibal morphs. Until recently, the prime candidates were a pair of microscopic fairy shrimp species (order Anostraca). David Pfennig and his colleagues proposed that the consumption of the shrimp by some of the spadefoot tadpoles served to trigger the cascade of genetically controlled developmental changes that transformed the shrimp-munchers into outsized cannibals.

But what was it about eating fairy shrimp that set this transformation into motion? Pfennig hypothesized that iodine-containing compounds found in the shrimp might cause the activation of specific genes in the tadpoles, genes that weren't turned on in the individuals that didn't consume shrimp. The prime candidate for a trigger substance turns out to be thyroxin, a thyroid hormone whose functions include stimulating metabolism and promoting tissue growth. A new set of experiments, though, have shown that even tadpoles that weren't fed fairy shrimp could still undergo the transformation to cannibals, indicating that (at the very least) something besides thyroxin intake must initiate the changes.

"What if it's not what they're eating but the mechanism of chewing itself that serves as a trigger?" I made the suggestion while brainstorming the problem with biologist Ryan Martin. "What if chewing on something alive like a fairy shrimp, something larger or something that struggles when you clamp onto it, sets this developmental cascade into motion?"

Martin shot me a "not bad for a bat biologist" look. "Sounds like a good grad student project."

"Hey, it's all yours," I said with a laugh. We then set to work, drawing up an outline for a potential experiment to test the hypothesis.

Although the jury is still out on the stimulus for the spadefoot transformations, Pfennig and his coworkers previously worked on a completely different cannibalism-triggering stimulus in another amphibian. And this one happened to be one of North America's most spectacular species.

Tiger salamanders (*Ambystoma tigrinum*) are the largest salamanders in the United States, reaching lengths of up to 13 inches. These thick-bodied, sturdy-limbed urodelans are widespread across much of the country.[7] Their markings, yellow blotches against a

---

7 Urodela (Greek for "conspicuous tail") is the order containing approximately 655 salamanders, lizard-shaped amphibians generally found in moist terrestrial environments.

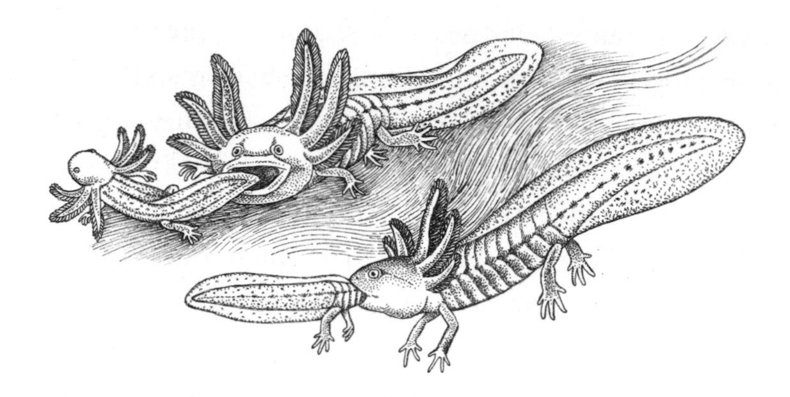

black body, make them easy to identify, but they are rarely seen in the open except during annual marches to a nuptial pond. Tiger salamander eggs are laid in the late winter or early spring, and like other salamanders (and their cousins the frogs and toads), their larvae are fully aquatic with external gills and fishlike tails. They typically feed on zooplankton and other micro-invertebrates, but under certain environmental conditions a small percentage of them develop traits that include huge heads, wide mouths, and elongated teeth. Consequently, these toothy individuals exploit larger prey, among them other tiger salamander larvae.

Pfennig and his colleagues set up lab experiments on fertilized *A. tigrinum* eggs to investigate the stimuli that set these changes into motion. First the researchers determined that the cannibal morphs only developed when larvae were placed into crowded conditions. Next, they used a variety of experiments to determine whether the larval transformation might be triggered by visual cues (that didn't work), smell (nope), or touch.

"It looks like they had to have the tactile cues," Pfennig told me. "There's something about bumping into each other that triggers the production of the cannibals."

*Movement and touch related*, I thought, remembering my suggestion about a possible trigger for the spadefoot cannibals. But instead of speculating about my own half-baked ideas, the conversation turned toward the pros and cons of cannibalism, especially as it pertained to consuming kin.

One of Gary Polis's general characteristics regarding cannibalism is that immature animals get eaten far more often than adults. Ultimately, this makes larvicide (or infanticide) the most common form of cannibalism in the animal kingdom. Intuitively, it doesn't seem logical to eat the next generation, but the behavior can make evolutionary sense for several reasons. Young animals not only provide a valuable source of nutrition, but in most species they're relatively defenseless. As such, they present instant nutritional benefits but little or no threat to larger members of the same species, most of which are invulnerable to attacks from immature forms.

But beyond acquiring a meal, and as we saw with spadefoot toads, cannibalism enables individuals of some species to accelerate their developmental process, thus allowing them to quickly outgrow a stage in which they might be preyed upon or perish due to unpredictable environmental conditions. In species like the flour beetle (*Tribolium castaneum*), the behavior may also impart a reproductive advantage, since studies have shown that cannibalistic individuals produce more eggs than non-cannibals. Finally, many animals maintain specific territories, within which they are intolerant to the presence of *conspecifics* (i.e., members of the same species). According to Polis, crowding increases the frequency with which individuals violate the space of others. By reducing overcrowded conditions, cannibalism can serve to decrease the frequency of territory violations.

There are also serious drawbacks to being a cannibal.

In all likelihood, the most significant of these is a heightened chance of acquiring harmful parasites or diseases from a conspecific. Both parasites and pathogens are often species-specific and many of them have evolved mechanisms to defeat their host's immune defenses. As a result, predators that consume their own kind run a greater risk of picking up a disease or a parasite than do predators that feed solely on other species. In the most famous example of cannibalism-related disease transmission, the Fore people of New Guinea were nearly driven to extinction as a result of their ritualized consumption of brains and other tissues cut from the bodies of their deceased kin, kin who had themselves been infected by kuru, an incurable and highly transmissible neurological disease. More on that topic later, but given its importance, the potential for disease transmission stands as a prime example that non-humans and humans alike share some of the negative consequences of cannibalism.

Cannibals—whether microbes or Methodists—who eat their own relatives can also experience decreases in a measure of evolutionary success known as *inclusive fitness*, in which the survival of an individual's genes, whether they're from an offspring or a collateral relative (like a brother or cousin) is the true measure of evolutionary success. A cannibal that consumes its own offspring, siblings, or even more distant relatives, removes those genes from the population, so it reduces its own inclusive fitness. Since this is bad juju, natural selection should favor cannibals that can discriminate between kin and non-kin, primarily because eating non-relatives results in no loss of inclusive fitness. In many instances, this is exactly what happens.

Because of the significance of inclusive fitness, it made perfect sense that David Pfennig and his colleagues had also worked on questions related to kin recognition, basically seeking to determine if some of their favorite cannibal species would avoid eating their own relatives. The researchers found that their study subjects did so by recognizing cues associated with their kin that were absent in non-kin.

"Most examples would fall under the heading of 'the armpit effect,'" Pfennig told me. "Here, an individual forms a template for what its kin smell like based on what its own smell is." He used the example of a species of paper wasps (subfamily Polistinae) that regularly raid the nests of conspecifics to provide food for their own broods. In these species, individuals learn that "If an individual smells like your nest or burrow . . . you don't eat them."

Similarly, tiger salamander larvae are more likely to eat the larvae of unrelated individuals than they are to consume relatives. Pfennig explained that he and his colleagues determined this experimentally by "preventing them from being able to smell."

"How did you do that?" I wondered, envisioning a team of micro-surgeons hovering over a tiny, amphibious patient. *Irrigation please, Nurse. Can't you see this patient is dehydrating?*

"By applying super glue under their nares," he replied.

"Oh, right," I said with an uncomfortable laugh, before Pfennig finished up by assuring me that the condition was temporary.

If you're wondering whether or not spadefoot toads avoid eating their kin, Pfennig told me that omnivores school preferentially with their siblings, whereas cannibals generally associate only with non-siblings. In close encounters of the bitey kind, cannibal tadpoles release siblings unharmed and consume non-relatives. In the lab, though, apparently all bets are off if the cannibals are deprived of food and then placed in a tank with other tadpoles. In these cases, starvation becomes the great equalizer, and both kin and non-kin are eaten. As I would learn from researchers unearthing new evidence about the Donner Party, this particular aspect of cannibalism spans the *entire* animal kingdom.

On the plane ride back to New York, I thought a great deal about the cannibalism I'd seen in the temporary ponds below the majestic Chiricahua Mountains, and about the tiger salamanders I'd collected at Long Island golf courses as a kid.

*Cannibal morphs.*

I wondered whether H. G. Wells knew about their existence when he wrote *The Time Machine* in 1895. In Wells's classic novel, the Time Traveler encounters two human species: the child-sized and docile Eloi, and the brutish Morlocks, who raise the Eloi in order to feed upon them. Wells explained the Morlocks' cannibalistic behavior by suggesting that they were once members of a worker class, toiling underground for lazy, upper-class surface-dwellers. The Time Traveler speculates that a food shortage (i.e., an environmental change) forced the subterraneans to alter their diets—at first rats, but ultimately something a bit larger. Eventually this behavior resulted in a race of hulking cannibals, feeding on the surface-dwellers, whose own evolutionary path would produce the sheeplike Eloi, pampered, well-fed, and eventually slaughtered for food.

Although the Eloi-Morlock relationship was clearly meant to serve as a cautionary tale of the horrors of class distinction, H.G. Wells imagined a biological phenomenon remarkably similar to what researchers like David Pfennig and his colleagues are working on today.

Many scientists now believe that phenotypic plasticity offers

the perfect building blocks for the type of evolutionary change described by Wells over a century ago. These building blocks could be novel traits like the tiger salamander's kin-chomping jaws or the spadefoot tadpole's serrated beak—each having originated as an environment-dependent alternative to an already established ancestral trait (in this case, normal jaws). What these scientists hypothesize goes far beyond the realm of cannibalism and into the very mechanisms of evolution itself. Their claim is that the appearance of new traits in a population, generally regarded as a first step toward the evolution of new species, can occur by means other than the accumulation of micromutations (i.e., small-scale or highly localized mutations), the classic mechanism by which new traits, and eventually new species, are thought to appear. Some researchers now believe that given generations, novel traits originating as examples of phenotypic plasticity have the potential to produce separate species.

This idea originated with the German-American geneticist, Richard Goldschmidt (1878–1958), infamous for his stance that micromutations accumulating over long periods of time were inadequate to explain the evolution of different species. He proposed two additional mechanisms, the first: speciation by macromutations (i.e., those causing a profound effect on the organism), which eventually led to the derision associated with his name and the legacy-destroying label of "non-Darwinian." Less well known is Goldschmidt's suggestion (quite correct, it appears) that mutations can result in major changes during early development, and that these can lead to large effects in the adult phenotype. This hypothesis and the related concept of developmental plasticity (i.e., adaptability) are two of the key principles of the modern field of evolutionary developmental biology (a.k.a. evo devo). Goldschmidt's contribution, though, is generally ignored. Along with the even earlier evolutionary biologist Jean-Baptiste Lamarck (the giraffe neck guy whose story is discussed in an upcoming chapter), these scientists are rarely celebrated for what they got right, and are, instead, derided for what they got wrong.

Okay, so now that I'd captured and examined cannibalistic tadpole morphs and heard all about their outsized salamander cousins, it was time to look into other examples of cannibalism in nature and to determine why these creatures were eating each other. I decided that the best way to cover and divvy up the material was to look at what I considered to be the most dramatic examples of Gary Polis's cannibalism-related generalizations. Admittedly, some of what I uncovered was hard to categorize, thus leading me to the realization that cannibalism can extend far beyond the realm of generalization. I also learned that normal behavior or not, sometimes cannibalism in the animal kingdom can get downright weird.

# 2: Go on, Eat the Kids

3RD FISHERMAN: *I marvel how the fishes live in the sea.*
1ST FISHERMAN: *Why, as men do a-land; the great ones eat up the little ones.*

—William Shakespeare, *Pericles*, act 2, scene 1

Many invertebrates do not recognize individuals of their own kind as anything more than food, and so a significant amount of cannibalism takes place within invertebrate groups like mollusks (clams and their pals), insects, and arachnids (spiders and scorpions). Thousands of aquatic invertebrates like clams and corals have tiny, planktonic eggs and larvae, and these are often a major food source for the filter-feeding adults. Since the planktonic forms often belong to the same species as the adults feeding on them, by definition this makes filter-feeding a form of indiscriminate cannibalism.

Although both fertilized and unfertilized eggs are eaten by thousands of species, the practice of consuming conspecific eggs appears to have led to the evolution of an interesting take on the concept of the "kids' meal." As the name implies, trophic eggs, produced by some species of spiders, lady beetles, and snails, function solely as food. These prepackaged meals often outnumber the fertilized eggs in a given clutch—a fact exemplified by the results of an observational study on the rock snail (*Thais emarginata*). This species commonly lays a clutch of around 500 eggs but averages only 16 egg-munching hatchlings.

The black lace-weaver spider (*Amaurobius ferox*) behaves similarly: one day after their spiderlings hatch, new mothers lay a clutch of trophic eggs, which are doled out to their hungry babies. The trophic eggs last for three days, after which the spiderlings are ready for their next stage of development, but in this case, the "smaller gets eaten" rule gets turned on its head.

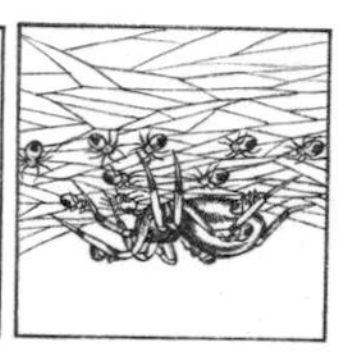

Arthropods like spiders, insects, and crabs are characterized by having their skeletons on the outside of their bodies. To grow in size, they undergo a regular series of molts, during which their jointed cuticle or exoskeleton is shed and replaced by a new skeleton arising from beneath the old. After their first molt and after the trophic eggs have been consumed, black lace-weaver spiderlings are too large for their mother to care for, though they are in dire need of additional food. In an extreme act of parental care, she calls the babies to her by drumming on their web and presses her body down into the gathering crowd. The ravenous spiderlings swarm over their mother's body. Then they eat her alive, draining her bodily fluids and leaving behind a husklike corpse.

Insects undergoing pupation, the quiescent stage of metamorphosis associated with the production of a chrysalis or cocoon, are also vulnerable to attack from younger conspecifics. The ravenous larva of the elephant mosquito (*Toxorhynchites*) not only consumes conspecific pupae, but also embarks on a killing frenzy, slaying but not eating anything unlucky enough to cross its path. The reason for this butchery appears to be the elimination of any and all potential predators before the larva enters the helpless pupal stage itself.

In some snail species, cannibalistic young transform into vegetarian adults. In one food preference test, hatchlings from an herbivorous snail fed on conspecific eggs exclusively (even when offered lettuce); four-day-old individuals ate equal amounts of eggs and lettuce; and 16-day-old individuals preferred the veggies. When snails older than four weeks of age were denied the lettuce option, they starved to death, even in the presence of eggs. The reason for this gradual transition in feeding preference appears to be that these snails, like other herbivores (from termites to cows), require a gut full of symbiotic bacteria before they can digest plant material. Since newly hatched snails have no gut bacteria, they're compelled to consume material that can be digested, even if this turns out to be their own unhatched siblings.

Cannibalism occurs in every class of vertebrates, from fish to mammals. For researchers, factors like relatively larger body size and longer lifespans have made these backboned cannibals easier to study than invertebrates. As a result, previously unknown examples of this behavior are being revealed on an increasingly regular basis. Additionally, factors related to the increased size and longevity of vertebrates have facilitated the ability of scientists to determine and track kin relationships (have you tried tagging a mosquito?), leading to a greater understanding of the complexities of cannibalism-related behaviors. One such result has been the classification of distinct forms of cannibalism, such as *filial cannibalism* (eating one's own offspring) and *heterocannibalism* (eating unrelated conspecifics), both of which have become vital to the concept of cannibalism as normal behavior.

In mammals, filial cannibalism has been reported in rodents (like voles, mice, and wood rats), and lagomorphs (rabbits and their relatives), as well as shrews, moles, and hedgehogs (a.k.a. The Mammals Formerly Known as Insectivores).[8] These mammal moms sometimes eat their young to reduce litter size during periods when food is scarce. Cannibalism also occurs when litter size exceeds the number of available teats or when pups are deformed, weak, or dead.

In the fishes, by far the largest of the traditional vertebrate classes, individuals in every aquatic environment and at every developmental stage are ambushed, chased, snapped up, and gulped down on a scale unseen in terrestrial vertebrates. One reason that cannibalism occurs so frequently in fish may be the fact that the group as a whole has more in common with the invertebrates (where cannibalism is often the rule and not the exception) than do the other vertebrate classes (reptiles, birds, and mammals). Another way to consider this is to think of the class Pisces as a mosaic—composed of a suite of more recently evolved, vertebrate traits (like a vertebral column and larger brain) but still retaining some invertebrate characteristics. Here it's the production of high numbers of tiny offspring with less parental care, as well as a proclivity for consuming both eggs and young—even one's own.

---

8 Insectivora and other traditional vertebrate classes like fish, reptiles, and birds are no longer considered to be valid units of classification, either because scientists have determined that species included in them aren't as closely related as they once believed, or that species that should be in those classes aren't.

At its most extreme, reproductive success in many fish species depends on a romantic-sounding technique known as broadcast spawning, during which females can release millions of eggs, while males simultaneously release clouds of sperm (milt). The end result is that some of the eggs get fertilized. Conceptually, given our own reproductive behavior, one might be misled into thinking that broadcast spawning is an inefficient mating technique. The bottom line, though, is that it works, as do similar variations on this theme employed by many amphibians and invertebrate species alike. Such reproductive strategies are successful because the vast number of eggs released offsets the low probability that any single egg will develop into a mature individual. Along those lines, scientists estimate that for every million eggs produced by an Atlantic cod (*Gadus morhua*), approximately one egg will result in an adult fish. Partially compensating for these lottery-like odds is the fact that each female produces between four and ten million eggs in a single spawning. On a related note, while this is a remarkable number of potential offspring for most vertebrates, it's something akin to sexual dysfunction in the ocean sunfish (*Mola mola*), a strange-looking, open-ocean species that can broadcast 300 million eggs in a single spawning—a vertebrate record.

But it's not just the abundance of eggs and young that makes fish such a popular menu item for members of their own species. Many terrestrial vertebrates produce few or even a single offspring, and most of these newly born or newly hatched individuals already exhibit considerable body size. In many fish species, the extraordinary number of eggs produced imposes a limit on their size, and so a full-grown cod might be a million times (or six orders of magnitude) larger than its own eggs. This fact goes a long way to explain why the majority of them exhibit about as much individual recognition of their offspring as humans do for a handful of raisins. Fish eggs, larvae, and fry (i.e., young fish) are vast in number, minute in size, and high in nutritional value. This makes them an abundant, nonthreatening, and easily collected food source. It's also why ichthyologists consider the absence of cannibalism in fishes, rather than its presence, to be the exceptional case.

Although not quite as infrequently practiced as it is among the invertebrates, parental care occurs in only around 20 percent of the 420 families of bony fishes (a group composed of nearly all living species except sharks and their flattened relatives, the skates and rays). The primary reason for this trend can be explained by the fact that the natural world is full of tradeoffs. Here the tradeoff works like this: Since females expend a tremendous amount of energy producing eggs (sometimes millions of them), they can't afford to expend much energy caring for them or their young when they hatch. For this reason, the eggs and fry of most fish species exist in dangerous environments inhabited by a long list of potential predators, including conspecifics. But even in the 90 or so piscine families where parental care does occur, filial cannibalism is an extremely common practice, and here the primary reason has to do with who is doing the babysitting.

Among the land-dwelling vertebrates, females are the principal caregivers, while males take on support roles or simply make themselves scarce. In bony fishes that guard their own eggs, though, it's usually the males who are involved, undertaking these chores at nests otherwise known as *oviposition sites*. These can range from slight depressions in the substrate, to rocks, plants, and other materials to which the sticky eggs (generally numbering in the hundreds) adhere in discrete clumps. The male guardians often wind up consuming some of the eggs (partial filial cannibalism), and sometimes all of them (total filial cannibalism).

One reason that male fish engage in this seemingly counterproductive behavior is that generally, they have much less invested in the brood than do females. It is less costly to produce a cloud of sperm than it is to produce, carry around, and distribute an abdomen full of eggs. Furthermore, with their ability to search for food seriously constrained by caregiving duties, males are forced to undertake at least some degree of fasting. This practice decreases their overall physical condition and thus the likelihood of future reproductive success. By consuming a portion of their own brood, males can increase the chances that they'll survive and produce additional offspring. New eggs are consumed more often than older eggs because there has been less parental investment in maintaining them.

In some examples, though, the loss of eggs from an oviposition site is not the fault of a hungry male guardian. Unrelated conspecific males regularly raid nests in order to consume or steal eggs. Egg theft can be explained by the preference for some females to spawn at sites already containing eggs, even if they're not hers. In these instances, once a female deposits her own clutch, the male will selectively eat the eggs he previously stole and deposited there.

While we're on the topic of parental care in fishes, mouthbrooding cichlids deserve a brief mention, if only because they serve to strengthen the often tenuous link between a mouthful of kids and lunch. Mouthbrooding occurs in at least nine piscine families, most famously in the freshwater Cichlidae. Cichlids, especially the African varieties, are extremely popular with aquarium keepers, as well as connoisseurs of tilapia—the Spam of gourmet fish. With more than 1,300 species, cichlids have evolved extremely specialized lifestyles that serve to reduce competition with related species living in the same area.

Mouthbrooding is a common form of behavior in cichlids. Typically, it refers to post-spawning behavior in which parents (usually females) hold their brood of fertilized eggs inside their mouths until they hatch and sometimes even after that. This provides the eggs and fry with a haven from predators, a point commonly portrayed in crowd-pleasing nature videos that depict young fish darting back into their parent's mouth at the first sign of danger. Conspicuously missing from these lighthearted reports is the fact that parents holding a mouthful of eggs usually eat a considerable portion of them, and sometimes the entire brood. Also destined for the digital equivalent of the cutting room floor are shots showing male cichlids fertilizing the eggs in the females' mouths, always a difficult topic to explain during family TV time.

Mouthbrooders practice filial cannibalism primarily because, as we all know, eating a regular meal is next to impossible while carrying around a mouthful of eggs. Cichlids and other mouthbrooders get around this vexing problem in the simplest way possible: cannibalism. Interestingly, scientists had thought that for the first few days after spawning, female mouthbrooders selectively consumed only unfertilized eggs from their broods. When researchers set out to determine just how mothers were able to distinguish between fertilized and unfertilized eggs, they were surprised to find that 15 percent of the consumed eggs were actually fertile. We now know that, mistakes aside, once all of the unfertilized eggs have been eaten, hungry mothers continue to consume small quantities of their own fertilized eggs. And should the brood reach about 20 percent of its original number, many mouthbrooders will write off the entire batch, and eat them all. As with similar examples of total filial cannibalism, this usually occurs when the cost of caring for the brood becomes higher than the benefit of producing a less-than-normal number of offspring. Rather than investing in a smaller brood, it becomes more advantageous for the female to recover some energy by consuming her remaining young and then moving on to find a new mate.

My personal favorite example of piscine cannibalism is yet another instance in which immature animals are the ones getting consumed. But in sand tiger sharks (*Carcharias taurus*), the individuals doing the cannibalizing haven't even been born yet.

Sand tigers, like hammerheads (*Sphyrna zygaena*) and blue sharks (*Prionace glauca*), do not deposit their eggs into the environment. Instead the eggs and young develop inside the females' oviducts, a developmental strategy known as *histotrophic viviparity*. Scientists who first looked at late-term sand tiger embryos in 1948 noticed that these specimens were anatomically well developed, with a mouthful of sharp teeth—a point (or several) driven home when one researcher was bitten on the hand while probing the oviduct of a pregnant specimen. Strangely, these late-term embryos also had swollen bellies, which were initially thought to be yolk sacs—a form of stored food. This was puzzling, though, since most of the nutrient-rich yolk should have been used up by this late stage of development. Further investigation showed that the abdominal bumps weren't yolk sacs at all, they were stomachs full of smaller sharks! These embryos (averaging 19 in number) had fallen victim to the ultimate in sibling rivalry—a form of *in utero* cannibalism

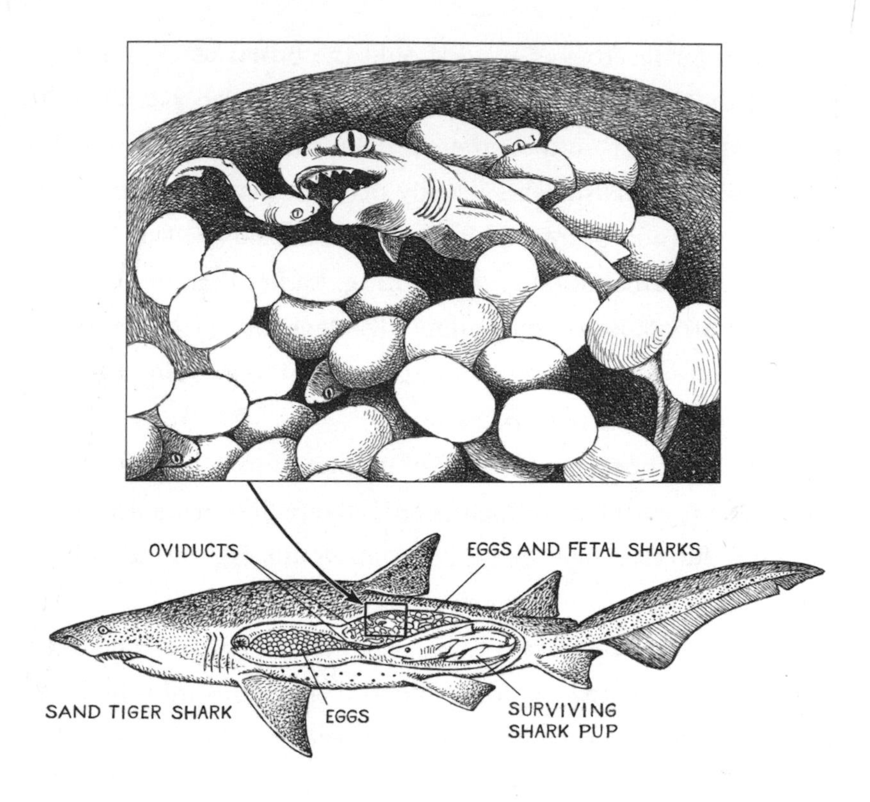


known as *adelphophagy* (from the Ancient Greek for "brother eating"), or sibling cannibalism.

This behavior is possible because sand tiger shark oviducts contain embryos at different developmental stages (a characteristic that also evolved in birds). Once the largest of the shark embryos run through their own yolky food supply, they begin consuming eggs. And when the eggs are gone, the ravenous fetal sharks begin consuming their smaller siblings. Ultimately, only two pups remain, one in each oviduct. According to renowned ichthyologist Stewart Springer, the selective advantage for the young sharks may extend beyond the obvious nutritional reward.

Springer, the first to study the behavior, believed that the surviving pups were born "experienced young," having already killed for survival even before their birth. He hypothesized that this form of sibling cannibalism might afford the young sand tigers a competitive advantage during interactions with other predatory species also looking for meal.

Although the sand tiger is the only species known to consume embryos *in utero*, several other sharks exhibit a form of *oophagy*, in which the unborn residents of the oviduct feed on a steady supply of unfertilized eggs. Additionally, a form of adelphophagy occurs in some bony fishes (superclass Osteichthyes) in which broods mature at different rates. Once again, in these species it's the older members of the brood that cannibalize their smaller siblings.

Cannibalism of the young also occurs in many species of snakes, lizards, and crocodilians, where, for example, it accounts for significant juvenile mortality in the American alligator (*Alligator mississippiensis*). Although reptiles do not transition through larval stages like most fish and amphibians, the smallest and most defenseless individuals, namely eggs, neonates, and juveniles, run the greatest risk of being eaten by conspecifics.

Cannibalism is relatively rare among birds, a fact that may be related to one aspect of their specialized anatomy—their beaks. These keratinous structures are responsible for the designation of most bird species as "gape-limited predators." In other words, their lack of teeth limits them to consuming prey small enough to be swallowed whole. Existing under this anatomical constraint, when cannibalism does occur in birds it's generally the eggs and young that are consumed.

According to Cornell ornithologist Walter Koenig, "Since brood reduction is widespread in birds, it's likely that sibling cannibalism would be a lot more widespread than it is if birds had beaks that were capable of tearing dying offspring to pieces, or opening their gape wide enough to swallow them whole."

Heterocannibalism (in which non-kin get eaten) has been reported in seven of the 142 bird families and is most common in colonial sea birds like gulls. Here, the practice of consuming eggs or young is an integral part of foraging strategy and it can have a significant effect on bird populations. In one study of a colony of 900 herring gulls (*Larus argentatus*), approximately one-quarter of the eggs and chicks were cannibalized. Heterocannibalism also occurs in acorn woodpeckers (*Melanerpes formicivorus*). In this species, two female woodpeckers share a single nest and will even feed and care for each other's young. But before that occurs, the nestmates may also destroy and consume each other's eggs if one bird should lay an egg first. Presumably this is because the oldest hatchling would be the most likely to survive. To eliminate this advantage, the birds will keep eating each other's eggs until they both lay their eggs on the same day, a process that can take weeks.

Sibling cannibalism, in which brothers and sisters get eaten by other brothers and sisters, is best known among the raptors, a non-taxonomic name for predatory birds like eagles, hawks, kestrels, and owls, all of which possess strong eyesight, powerful beaks, and sharp talons. The latter two characteristics exempt raptors from the gape limitations seen in many other birds and may help explain the increased frequency of cannibalism in these birds. In some species, sibling cannibalism is often the end result of *asynchronous hatching*, in which two eggs are laid with one of them hatching several days before the other (one can almost hear the acorn woodpeckers gasping in horror). As a result, the firstborn chick uses its extra bulk to win squabbles over food with its younger brother or sister. In instances where the parents are unable to provide their young with enough to eat, the firstborn will kill and consume its younger sibling. Researchers sometimes refer to these types of victims as "food caches," as sibling cannibalism becomes an efficient way to produce well-nourished offspring (albeit fewer of them) during times of stress.

Something similar happens in the snowy egret (*Leucophoyx thula*), which commonly lays three eggs. The first two get a serious dose of hormones while still in the mother's body. The third egg receives only half the hormone boost, resulting in a less aggressive hatchling. If food is abundant, the larger nestlings simply throw the passive chick out of the nest, but if alternative sources of nutrition become scarce, the smaller sibling is stabbed to death and eaten.

According to Koenig and fellow ornithologist Mark Stanback, filial cannibalism in birds has been reported in 13 of 142 avian families but is not well understood, perhaps because it is relatively infrequently observed. On rare occasions, birds like roadrunners (*Geococcyx californianus*) will eat undersized chicks. Similarly, barn owls (*Tyto alba*) are reported to consume their own chicks during extreme environmental conditions, such as when food is in short supply and the chicks are either starving or sick. It has been suggested that filial cannibalism of dying or decayed offspring can prevent infection and deterioration of the entire clutch. Presumably there are also benefits to getting rid of dead chicks before they attract legions of carrion-eating flies and maggots. In most cases, however, it's the lack of alternative forms of nutrition that initiates the behavior.

"A lot of examples of cannibalism in birds are clearly associated with food limitation," Koenig told me. "We're basically talking about a lifeboat strategy, where the strong cannibalize the weak."

# 3: Sexual Cannibalism, or Size Matters

*Whoever authorized the evolution of the spiders of Australia should be summarily dragged out into the street and shot.*

—Mira Grant, *How Green This Land, How Blue This Sea*

While it's fairly common knowledge that the praying mantis is the co-holder (along with the black widow spider) of the title "Nature's Most Infamous Cannibal," fewer people know that the name *praying mantis* is shared by nearly all of the 2,200 species making up the order Mantodea.

The moniker comes from the curious manner in which the insects hold their forelegs while resting. As a result of this prayer-like attitude, they've become some of the most popular insects in mythology and folklore. Many of these mantid myths have religious or semi-religious overtones. In France, *par exemple*, they're known as *prie-dieu* and are said to point lost children homeward. The Khoi people of South Africa regard praying mantises as gods, while Arab and Turkish folklore holds that the insects direct their prayers toward Mecca. Always looking on the bright side, Americans once believed that praying mantises blinded people and killed horses, this perhaps as a nod to the fact that rather than being used for prayer, the anterior-most limbs are actually modified into lethal, spike-covered weapons. Often well camouflaged, most species are ambush predators, lashing out with their "raptorial legs" to capture, crush, and secure their prey while a set of sharpened mouth parts slice, can-opener style, through the toughest exoskeleton.

Although mantises feed primarily on other insects, the largest species can reach around six inches in length and these giants will attack and consume small reptiles, birds, and even mammals. It is likely that this type of predatory behavior is responsible for the common misspelling "preying mantis."

As a child in the 1960s I was told that there would be a $50 fine for anyone caught killing a praying mantis (and my friends have the same recollections). Since I was unable to uncover a record of any such federal or state law, I can only assume that the story was a scare tactic designed to keep nasty little boys from slaughtering an uncommonly pious insect known to eliminate an array of less religiously inclined pests.

Many people are familiar with the praying mantis's supposed penchant for cannibalistic sexual encounters, reports of which began showing up in the scientific literature in the late 19th century. Back then, several authors claimed that female mantises regularly bit off the triangular heads of their partners during sex. These same sources also claimed (to many a reader's astonishment) that the decapitated males continued to copulate, abdomens pulsing away as if nothing of much importance had just happened. According to these references, several hours later the female would stride off,

full and fertilized, while the male, having been reduced to a tiny pile of wings and hard bits, stayed put. Similar tales about mantid mating continued into the early 20th century, when members of a new generation of entomologists began investigating the function of this rather puzzling behavior.

One hypothesis reasoned that the male mantis's brain actually inhibited sexual behavior. With their heads removed, however, males became "disinhibited," found the rhythm, and eventually pumped out a full load of sperm. Other mantid mavens suggested that getting oneself cannibalized made sense for praying mantis males that might have limited opportunities to mate over their lifetime. It made evolutionary sense, therefore, to fatten up the only female they might ever run into—especially one now carrying their sperm. Furthermore, it would be a plus for both sexes since headless males reportedly pumped out more sperm than those equipped with heads, leading to more fertilized eggs and more offspring. These accounts contributed to an overall impression that the decapitation of male mantises was a normal and perhaps necessary copulatory stage and, soon after, the concept became entrenched in textbooks and the popular literature. Unfortunately, what never quite made it into print was the fact that most observations of mantis cannibalism were made in laboratory settings and only after females had been deprived of food.

In reality, cannibalism varies across this large and diverse group. The behavior has gone unobserved in most species, not necessarily because it doesn't happen, but because it hasn't been studied. Researchers now believe that rather than being a required component of mating behavior, the consumption of males is more likely to be a foraging strategy employed by hungry females unable to wrap their raptorial forelegs around an alternate form of nutrition.

Support for this hypothesis comes from studies on a wide variety of mantis species, including those in which worse-for-wear females that cannibalized their mates later exhibited improved body condition, produced larger egg cases (*ootheca*) and more offspring. Significantly, well-fed female mantises showed no cannibalistic tendencies during mating encounters.

Before we blame mantid cannibalism on captive conditions or starvation, though, the fact remains that both wild and captive males exhibit extreme caution as a normal preamble to copulation. Depending on the species, the males' initial approach can vary from simple (slow and deliberate movement toward the female, followed by a flying leap onto her back) to complex (the male fixes its stare on the female, goes through a series of stereotypical movements like antennal oscillations and abdominal flexing, *then* takes a flying leap onto her back). Researchers believe that these movements serve to either circumvent or inhibit the females' aggressive, predatory response. It is, therefore, extremely unlikely that these forms of cautionary behavior by males would have ever evolved if there weren't at least some risk of being attacked by females.

And what about the male's famous ability to "keep the beat" even after losing its head? Biologists Eckehard Liske and W. Jackson Davis have an explanation for that phenomenon as well. They believe that, rather than acting as a stimulus for copulation (by releasing sexual movements), decapitation artificially induces the behavior. This would be similar to the way in which lopping off a chicken's head artificially induces locomotor movements that can temporarily propel a headless bird around a barnyard. According to these researchers, from an evolutionary perspective, reflexive abdominal contractions and the subsequent release of sperm may insure that fertilization takes place, even if the males are left feeling a bit lightheaded after sex. As such, it serves as a prime example of how cannibalism can benefit the individual being cannibalized.

As in the praying mantises, in certain well-known spiders, truth has been masked by myth. After several papers in the 1930s and 1940s reported that female *Latrodectus mactans* spiders devoured their mates after copulation, *L. mactans* and two additional North American species became widely known as black widows. Although most of the initial observations turned out to be anecdotal, cannibalism and black widows became forever linked, appearing in an array of literature that ranged from storybooks to college textbooks on evolution, ecology, and animal behavior. The cannibal association continued through the 1970s and 1980s, even though researchers working with these spiders were beginning to discover that the behavior in black widows was actually a rare occurrence. They determined that not only did most male spiders depart unharmed after copulation, but some of them lived in the female's web for several weeks, even sharing her prey.

"The supposed aggressiveness of the female spider toward the male is largely a myth," said spider expert Rainer Foelix. "When a female is ready for mating, there is little danger for the male." Foelix did add that all bets were off if a male mistakenly showed up in the web of a hungry female.

Readers who might be disappointed to learn that the black widow's reputation is apparently worse than its bite may be consoled by the fact that sexual cannibalism has been reported in 16 out of 109 spider families (although the list is described as not "exhaustive nor definitive" regarding frequency). One of the most interesting examples takes place in the black widow's Aussie cousin, the redback spider (*L. hasselti*). In this species, males go to extreme lengths, not only to guarantee their own demise, but their consumption as well.

The Australian redback spider is common throughout Australia, and in a country renowned for its notorious creatures, the redback ranks among the most dangerous. The reasons behind this spider's bad reputation start with a neurotoxic bite that can cause severe pain and swelling, and in rare instances, seizures, coma, and even death. Like the North American black widows, Australian redbacks are often found in close proximity to human residences, especially sheds and garages offering the spiders undisturbed areas full of clutter. Presumably because of the abundance of flies, both black widows and redback spiders were once common in outhouses, where their fondness for living under privy seats was never quite as unpopular as their habit of biting anything that blocked their escape routes. Slim Newton's song "Redback on the Toilet Seat" details one such encounter. "There was a redback on the toilet seat when I was there last night—I didn't see him in the dark but boy I felt his bite." The song reached #3 on the Australian pop record charts in August 1972, well ahead of artists like The Bee Gees and Elton John.

Although encounters with humans are rarely fatal (at least for the human), the same cannot be said for male redback spiders attempting to mate. In the first stage of courtship, the male approaches the female's web and proceeds to get himself noticed (which takes a bit of doing, since he is only about one-fifth her size). He does so by bouncing his body up and down, throwing some silk around, and waving his front legs. As a point of information, while insects have six legs, spiders have eight of them, plus an additional pair of anteriorly located appendages called *pedipalps*. In male spiders, the pedipalps are modified for transferring sperm to the female's body, a chore necessitated by the fact that spiders lack penises. Furthermore, there is no internal connection between the pedipalps and the testes, which are located within the abdomen. Instead, sperm is initially extruded from a furrow on the male's abdomen into a spun receptacle called a sperm web. As a male dips his pedipalps into the pooled sperm, a pair of coiled structures called *emboli* and their associated muscles work like tiny turkey basters to suck up the liquid and store it until copulation.

The next phase of redback courtship begins as the male initiates repeated bouts of physical contact with his potential mate, behavior that includes tapping, probing, and nuzzling. The real heavy

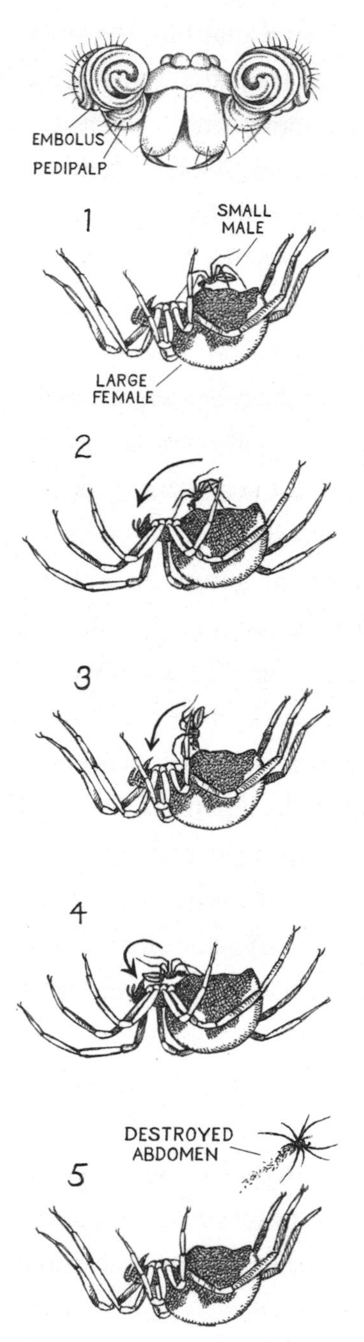


petting begins once the male locates the female's epigynal opening. According to spider expert L.M. Forster, this behavior includes "nibbling, palpal boxing, and knocking, embolus stretching, push-ups, abdominal vibration, and epigynal scraping."

By now, if the female hasn't already eaten the male (which can put a serious dent into all of this foreplay) the spiders briefly assume "Gerhardt's position 3." To visualize this, picture two people in the missionary position. Now tweak the imagery a bit so that the guy is approximately the size of your favorite throw pillow. Okay, now add another eight limbs. (All right, maybe you shouldn't picture this.)

Gerhardt's position 3 appears to be favored by all *Latrodectus* species except the Australian redback, where it is abandoned immediately after the male penetrates the female's epigynal opening with the tip of a sperm-charged embolus. At that point, things take a hard turn toward the strange. The male slowly performs a 180-degree somersault that ends with his abdomen resting against the female's mouthparts and she immediately expresses her gratitude by vomiting enzyme-laden gut juice onto the tiny acrobat. She then begins to consume the male's abdomen as they copulate, pausing from time to time to spit out small blobs of white matter. Upon the completion of the sex act, which takes anywhere from five to 30 minutes, the male crawls off a short distance, reportedly making repeated attempts to reel in his spent embolus "by stretching it with his forelegs and then releasing it abruptly."

Approximately ten minutes later, rather than fretting over missing body parts or a less-than-tightly-coiled sperm applicator, the male returns to the fray, this time wielding the second embolus. The half-eaten spider then proceeds to reenact its earlier copulatory acrobatics, although this time minus some viscera. By way of a "welcome back," the female resumes her meal, consuming more and more of the male's abdomen. At the end of this round, though, rather than allowing him to crawl off, the female wraps her shredded partner in silk, eventually snorking up his now enzyme-liquefied innards like a spider-flavored Slurpee.

While the benefits of a risk-free meal for the redback mom-to-be are fairly obvious, one has to wonder what the hell is in it for the male? Because of this very question, the mating habits of *L. hasselti* have drawn interest from spider experts. The puzzled scientists determined that females that had recently eaten their mates were less receptive to the approach of subsequent suitors. Cannibalized males also copulated longer and fathered more offspring than non-cannibalized males. Ultimately, then, it seems that this rather extreme example of paternal investment optimizes the likelihood that the cannibalized dad gets to pass his genes on to a new generation.

Things get dicey, though, when trying to determine the benefits for redback males eaten before mating takes place, a situation that has also been reported in orb-weaving spiders like *Araneus diadematus*. Mark Elgar and zoologist David Nash worked with this species and proposed that pre-mating cannibalism allows the female to choose which male will get to inseminate her. The researchers supported their hypothesis with the observation that smaller males were eaten more often than larger—and presumably healthier—individuals. They also used modeling studies to hypothesize that pre-mating cannibalism would occur only in instances where there was no shortage of males from which to choose. Alternately, post-mating cannibalism appears to make good evolutionary sense when short-lived males are relatively few in number and have a low probability of encountering receptive females. In these instances, when a potential mate is encountered, it pays to give it your all, even if that means paying with your life.

Observations related to mantis and spider cannibalism serve to illustrate the Gary Polis generalization that among invertebrate cannibals, males get cannibalized far more frequently than females. This behavior occurs in species that exhibit *sexual dimorphism*, a condition in which there are anatomical differences between males and females of the same species.

Besides body size, other examples of sexual dimorphism include coloration and ornamentation, and here it's usually the males that display bright colors, or exhibit elaborate structures like horns, frills, and crests. Essentially, these showy accouterments are used to advertise the wares of a presumably healthy male to potential mates (in other words: "If I can spend energy advertising just how beautiful I am, then I'm also a great bet to father a clutch of healthy offspring").

As it relates to cannibalism, the most common example of sexual dimorphism is body size, and among the invertebrates, females are often substantially larger than males. But why is this so?

The benefits appear related to the ability of larger mothers to better carry, protect, and provide for their young. Relatively small body size can also provide males with a gravity-related biomechanical advantage. Since gravity is less of a constraint on lightweight bodies than it is on heavier ones, this is especially evident in males that must climb in order to reach females. Likewise, in other species, tiny adult males are able to "balloon" like spiderlings, using the wind as an energy-efficient means of travel. Presumably this increases their chances of finding a potential mate.

On the other side of the web, but still supporting the small-get-cannibalized rule, are the rare instances in which male spiders are larger than females. This spider role reversal occurs in two species that exhibit some very unspiderlike behavior. In the sand dwelling wolf spiders (*Allocosa brasiliensis*), females undertake risky visits to burrows built by the larger males. Because these structures represent a high reproductive investment, male wolf spiders become extremely picky when females show up and initiate courtship—which they do by alternately waving their forelegs around in the universal signal for "Pick me! Pick me!" In many cases, though, researchers noted that instead of mating, the females were often attacked and cannibalized.

To determine why, arachnologist Anita Aisenberg and her colleagues performed experiments in which twenty male spiders were consecutively exposed to one virgin and one previously mated female (in alternating order). Findings revealed that only 10 percent of the virgin females were cannibalized while 25 percent of the mated females were eaten, especially those exhibiting lower body condition indices. In other words, male wolf spiders chose their mates based on looks and sexual history. The researchers concluded that by selecting younger, fitter females, male spiders maximized the likelihood that their mate would survive to produce a successful batch of spiderlings. Older, less fit females, also served a purpose—as food.

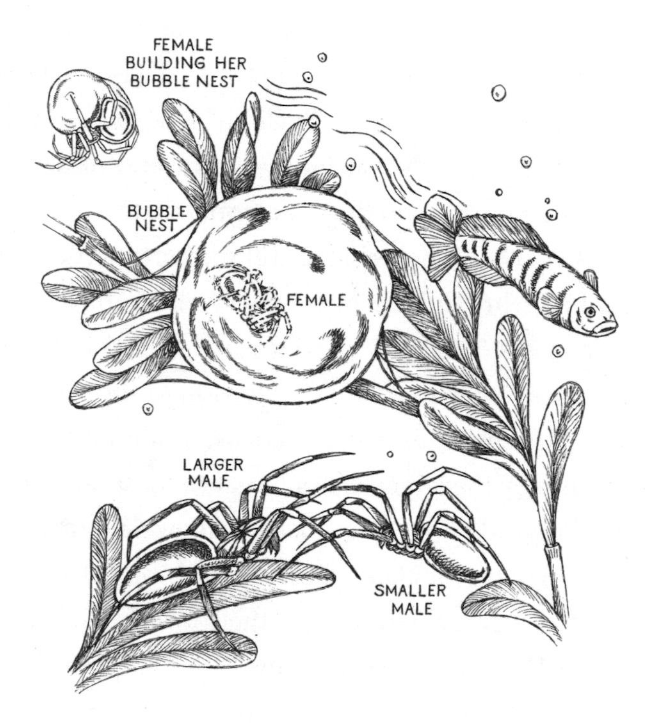


Cannibalism by males also occurs in the aptly named water spiders (*Argyroneta aquatica*), the only living arachnids that exist completely underwater. In this species, females spend most of their lives inside web-shrouded air bells, where their smaller bodies require less oxygen than their male counterparts. Natural selection may favor larger body size in male water spiders by providing them with enhanced swimming and diving abilities. While females are ambush predators, males are active hunters, and although their diets consist primarily of insect larvae, they will kill and cannibalize smaller males during intense competition for females. The female water spiders' preference for larger males can also turn deadly in a hurry, specifically during failed mating encounters in which females are chased down and consumed.

These two examples illustrate that when cannibalism occurs, it's size, rather than sex, that is the key determinant, with the smallest individuals (usually the males) ending up on the menu. This generalization also extends from invertebrates like spiders to some of the most familiar and beloved creatures on the planet. But before you come away thinking that praying mantises and spiders have maintained their stranglehold on cannibalistic copulatory behavior, we need to drop in on yet another member of the animal kingdom.

WHEN TERRESTRIAL SNAILS cross paths (or more accurately, slime trails), the potential for bizarre sexual encounters can rival a bachelor party in a Hangover film. For the snails, the high hook-up ratio stems from the fact that most of the participants are simultaneous hermaphrodites, enabling them to exchange sperm while at the same time having their own eggs fertilized. And while this particular sexual orientation increases the likelihood that any two individuals that meet will be able to mate, things can go downhill quickly once the lovers begin biting chunks out of each other.

Snails and slugs, their shell-challenged relatives, are mollusks, a biologically diverse invertebrate group that also contains the bivalves (clams, oysters, and their shelly relatives) and cephalopods (squid, octopuses, and cuttlefish). Known collectively as gastropods, the approximately 85,000 species of snails and slugs have a worldwide distribution, inhabiting a variety of marine, freshwater, and terrestrial environments. To put this into perspective, there are approximately 17 times as many gastropod species on the planet as there are mammals.

Along with their popularity as *escargots* or *scungilli*, gastropods are renowned for their slow-footed locomotion—a point celebrated annually by pub-going "researchers" in the UK. At the World Snail Racing Championship in 1995, a garden snail named Archie rocketed across a 13-inch course at an average speed of 0.0053 miles per hour, the fastest ever recorded for a snail. Scientists believe that snail speed (or lack thereof) is actually an adaptation related to energy efficiency. In other words, by devoting less energy to locomotion, gastropods can spend more of it involved in alternative behavior—like mating.

Although snail sex can last for up to six hours in some herbivorous species, this is definitely not the case in certain carnivorous gastropods, where foreplay can turn into cannibalism in the blink of a turreted eye. In these species, since even copulating individuals will bite their mates, each potential sexual partner is also a potential predator. As a result, they often employ the wham-bam-scram approach during sexual encounters, which can sometimes linger on for as long as six seconds.

An even more cringeworthy behavior is exhibited by banana slugs (genus *Ariolimax*), which become so entwined during sex that they sometimes chew off their partner's corkscrew-shaped penis in an effort to disengage. During this process, which is known as *apophallation*, penises are slurped down spaghetti-style, occasionally by their owners. Although this usually puts an end to the festivities, the fact that the penises do not grow back presents fewer problems than one would expect. The hermaphroditic slugs simply carry on the remainder of their lives as females.

In some land snails, however, things get bizarre even before the member-munching starts—especially once the partners begin shooting calcified "love darts" at each other, an exchange initiated when the body of one snail touches that of a potential mate. This tactile stimulation triggers the release of built-up hydraulic pressure in a sac surrounding the dart. As a result, the barbed projectile (known as a *gypsobelum*) explodes outward, embedding itself in the body wall of the second individual. In most instances, the skewered snail responds by shooting a dart of its own, and shortly thereafter the couple appear to remember why they had gotten together in the first place.

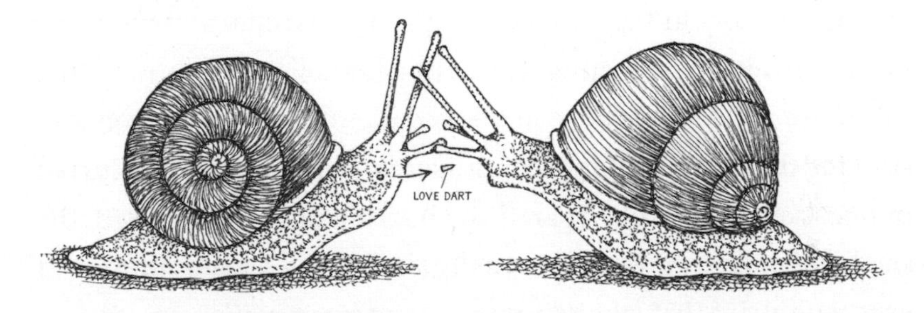


Often, though, the exchange proceeds with something less than textbook precision. Since most snails are nocturnal, their visual systems are simple. They can differentiate between light and dark, but an inability to determine details about their slimy targets (or anything else, for that matter) can lead to a serious lack of accuracy. As a result, headshots and similar misfirings are a common occurrence.

The obvious question is: Why do some snails fire miniature harpoons at each other? The proposed function of this behavior (which is rarely observed in humans anymore!) has undergone some revision. Earlier snail experts thought that love darts were the equivalent of an exchange of wedding gifts—in this instance, calcium carbonate, a major component of the snails' shell and eggs. Another suggestion was that the projectiles might act as an aphrodisiac or that they somehow signaled the shooter's willingness to mate. But support for these hypotheses never materialized.

I posed the question to McGill University biologist Ronald Chase, whose work in the 1990s helped solve the mystery of this baffling behavior. "The darts serve to increase paternity," he told me, since snails scoring love dart hits on their partners before mating fathered twice as many offspring as those that didn't hit their targets. The key to the enhanced reproductive effect was the tiny projectile's chemical coating. Chase and his colleagues showed that this hormonelike substance prevented digestive enzymes from destroying the majority of incoming sperm, something that occurred in non-skewered snails. Spared the buzzkill of being digested, the snail sperm sped onward, eventually fertilizing a greater number of eggs than those that wound up in non-speared snails.

Additionally, a 2013 study by Japanese researchers showed that snails skewered by love darts delayed re-mating with other individuals, an indication that something in the dart's mucous coating suppressed subsequent mating behavior—thus reducing the possibility that another male's sperm would outcompete the dart shooter's.

According to Chase, "It's all basically sexual selection." In other words, in any given population, some individuals outproduce other individuals because they're better at securing mates, usually by making themselves more attractive to the opposite sex or by beating back the competition. In land snails, explanations for who got the edge and how they achieved it are confounded by the fact that mating individuals not only exchange sperm with each other, but explosive projectiles as well.

Before leaving the topic of snails, if all this talk about love darts has you thinking about one of our most endearing holiday characters, you aren't alone. Ronald Chase believes that Cupid, the Roman version of the Ancient Greek god Eros, had his origin in land snails and their love darts.

"I think that the Cupid myth arose from Ancient Greeks observing snails mating and shooting love darts," Chase explained. "The species that we worked on in our experiments is found in Greece and I'm sure they shoot love darts over there as well." Ever the scientist, Chase added that there was no hard evidence yet and that neither he nor his students was able to find images of snails shooting love darts on Ancient Greek coins or pottery. Personally, I've always been a bit creeped out by the idea of a nude, weapon-wielding infant with wings, but considering that the Greeks could have equipped him with turret eyes and a slime trail, I'm willing to cut the current incarnation some slack.

# 4: Quit Crowding Me

*Hunger has its own logic.*

—Bertolt Brecht

Overcrowded conditions often coincide with another of Gary Polis's cannibalism-related generalizations, namely that incidents of cannibalism increase with hunger and with a decrease in the availability of alternative forms of nutrition, a point that will become horribly clear once we begin our investigation of human cannibalism.

Carrying the banner (albeit a tiny one) for crowd-related cannibalism are the Mormon crickets (*Anabrus simplex*). These insects are native to the North American West and belong to the order Orthoptera, which contains grasshoppers, crickets, and locusts. The fact that *A. simplex* is actually a form of jumbo katydid also makes them members of an unofficial assemblage composed of misnamed animals like "flying foxes" (which aren't foxes) and "tree shrews" (guess).[9] Attaining a body length of nearly three inches, Mormon crickets are

9 Biological nomenclature is full of misleading scientific names. *Vampyressa*, *Vampyrodes*, *Vampyrops* and *Vampyrum* are all bat genera, but none of them feed on blood. There are also bad puns, like *Apopyllus now* (a sac spider) and *Ittibittium* (a tiny mollusk), as well as rude sounding names, like *Pinus rigida* (the pitch pine) and *Enema pan* (a scarab beetle).

flightless, but like their winged cousins, the grasshoppers and locusts, they're renowned for their spectacular swarming behavior and mass migrations. According to biologist and Mormon cricket expert Stephen Simpson, favorable early spring conditions like warm weather and moisture can lead to the nearly simultaneous hatching of several million individuals. Almost immediately, the nymphs begin to march, and they do so in a spectacularly well-coordinated manner.

I asked Simpson why Mormon crickets participated in such large-scale movements. He cited studies showing that individuals separated from their swarm suffered 50 to 60 percent mortality from predators. "They got eaten by birds, rodents, and spiders if separated but were safe from predation in a crowd."

Seeking to illuminate principles of mass migration and collective behavior, Simpson and his coworkers conducted food preference tests on captive Mormon crickets. They determined that protein and salt were the limiting resources being sought by the swarming insect masses. Incidents of cannibalism began soon after these resources were depleted, since the nearest source of protein and salt becomes a neighboring cricket. According to Simpson, "Each insect chases the one in front, and in turn is chased by the cricket behind." Stopping to eat becomes a dangerous behavior, the biologist explained, requiring individuals to fend off other members of the swarm with their powerful hind legs. "Losing a leg is fatal," he told me. "The weak and the injured are most at risk."

Simpson demonstrated this experimentally by gluing tiny weights to some of the crickets, thus causing them to lag behind their unencumbered swarm-mates. Almost immediately, the miniature Jacob Marleys were attacked and eaten by the hungry horde approaching from behind.

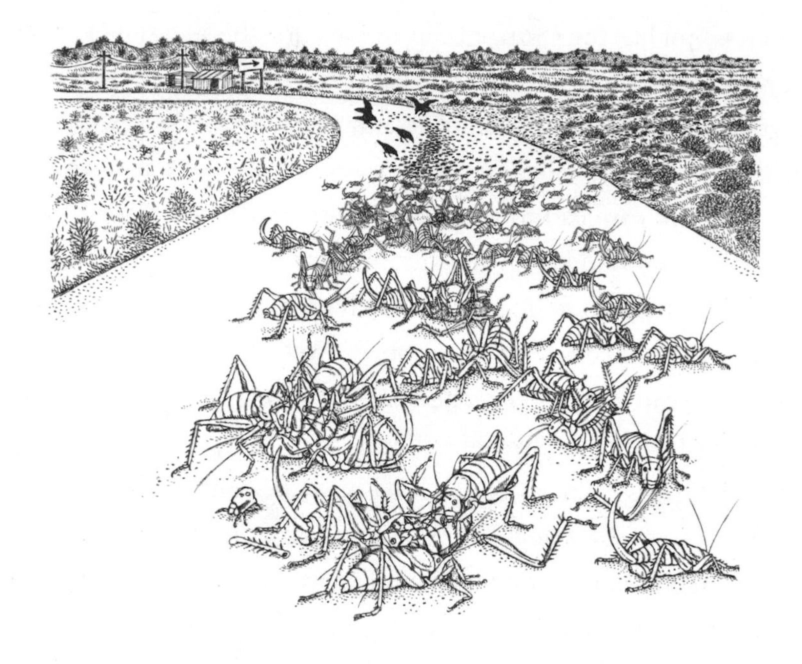

In the end, Simpson and his colleagues determined that the massive migratory bands were actually forced marches, demonstrating "coherent mass movement at the level of a huge marching band." Here, though, band members that can't handle the pace run a serious risk of being eaten.

WHILE AVIAN CANNIBALISM might be relatively rare in the wild, all bets are off once birds are removed from their natural setting, and packed shoulder-to-shoulder (or ruffled feather to ruffled feather). When thousands of stressed-out birds have little to occupy their time, the situation can deteriorate rapidly. In these instances the real meaning of the term "pecking order" becomes gruesomely apparent as some individuals are pecked to death and eaten. Initially, cannibalism on poultry farms was thought to result

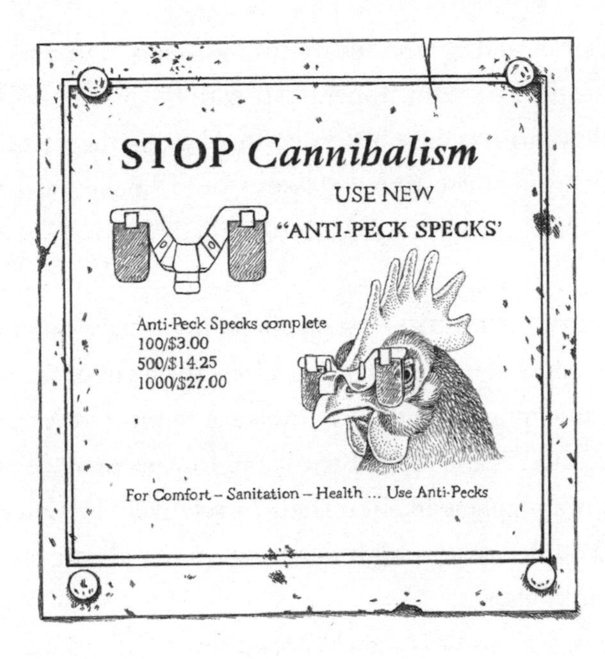


from a protein deficient diet, but researchers now believe that it's actually misdirected foraging behavior related to cramped and inadequate housing conditions.

As the poultry and egg industries became established, feather pecking and cannibalism (known in the trade as "pick out") became two of the most serious threats faced by poultry farmers. To stop cannibalism and prevent the loss of their valuable egg-laying hens, farmers routinely clipped off the tip of the bird's beak, a reportedly painful process. In the 1940s, however, the National Band and Tag Company came up with a far more painless and fashion conscious method to deal with the problem of cannibalistic chickens. Their design team reasoned that if the birds couldn't see "raw flesh or blood" then they wouldn't cannibalize each other and so they came up with "Anti-pix"—mini sunglasses equipped with red celluloid lenses and aluminum frames. Purchased in bulk ($27 for 1,000) and attached to the upper portion of the bird's beak near the base, poultry farmers were informed that having their chickens see the world through rose-tinted glasses would "make a sissy of your toughest birds," and apparently they worked.[10]

CURRENTLY, ONLY 75 species of mammals (out of roughly 5,700) are reported to regularly practice some form of cannibalism. Although this number will likely increase as more researchers become interested in the topic, the overall low occurrence of cannibalism in mammals is likely related to relatively low numbers of offspring coupled with a high degree of parental care (compared to non-mammals).

The golden hamster (*Mesocricetus auratus*), also known as the Syrian hamster, is a popular pet for children, but these cuddly fuzz balls are also known to display some nightmare-inducing behavior in captivity. The problems stem from major differences between their natural habitats and the captive conditions under which they are typically held. Native to northern Syria and Southern Turkey, *M. auratus* lives in dry desert environments. Adults are solitary, highly territorial, and widely dispersed. Individuals inhabit their own burrows and emerge for short periods at dawn and dusk to feed and mate. This crepuscular lifestyle is thought to help them avoid nocturnal predators like owls, foxes, and feral dogs. The

10 Although Anti-pix specks are now collector's items, the idea behind them lives on in plastic clips called "Peepers," which can be attached via a pin through the nostrils of various commercially raised game birds. For an extremely entertaining short on the original Anti-pix specks, check out the following link: http://www.nationalband.com/Chickenglasses.mov

results of a study on golden hamsters in the wild emphasized the major differences between natural conditions and those imposed on pet hamsters. For example, the researchers determined that in the wild, the average time hamsters spent on the surface during a 24-hour period was 87 minutes.

The problems between natural and captive conditions often begin in pet shops, where male and female golden hamsters are often kept in unnaturally large groups and displayed in well-lit aquaria. They are purchased singly (preferable) or in pairs (males and females if the store personnel know how to differentiate sexes). As pets, these desert-dwellers are housed in cages or trendy modular contraptions where translucent plastic tubes link "rooms" to each other. Unfortunately, the cages are often too small and golden hamsters have a hard time fitting through the plastic tubes, especially when pregnant or obese from overfeeding. Cage floors are usually covered in cedar shavings, which are pleasant enough to the human nose but hardly reminiscent of a desert environment. Regularly handled by children and often subjected to excessive noise and damp conditions (resulting from soiled cage bedding or leaky water bottles), many pet hamsters spend their existence under the watchful gaze of dogs and cats, their owners blissfully unaware that these are the hamster's natural enemies.

As a result of this laundry list of captivity-related stresses, female golden hamsters, especially younger ones, frequently cannibalize their own pups. Beyond diet (too much or too little food) and housing conditions, cannibalism can be triggered if hamsters are handled late in their pregnancy or if the babies are handled within ten days of their birth. The presence of additional individuals (even fathers) can also lead females to consume their own pups and heterocannibalism can occur if adult females encounter unrelated young. Pet experts suggest that filial cannibalism can be prevented by isolating pregnant individuals, making sure that water and nutritional requirements are met, and refraining from handling female hamsters before and after they give birth. Finally, pet owners should not handle newborns of *any* species unless they are prepared to nurse the animals themselves.

Cannibalism of adults can also take place when several mature golden hamsters are kept in the same cage, and this includes siblings, who reach sexual maturity at around four weeks of age. Under these conditions, fighting is common, and serious injuries or even fatalities can result. In the latter instances, the survivor of the battle typically consumes the carcass of the loser.

Although mice, rats, guinea pigs, and rabbits also occasionally cannibalize their young in captivity (primarily when food and water are scarce), there are several factors that appear to make golden hamsters even more prone to this type of behavior. Most significant is the fact that *M. auratus* has the shortest gestation period (16 days) of any placental mammal, and they can become pregnant again within a few days of giving birth. This means that females, already weakened and stressed out by the rigors of pregnancy, birth, and nursing, may be tending a new brood of eight to ten pups less than three weeks after their previous delivery.

When non-human primates (i.e., monkeys and apes) are compared to other mammal groups, cannibalism is rare, having been observed in only 11 of 418 extant species. Many examples of primate infanticide and/or cannibalism were thought to be stress-related and, generally, this turned out to be true. Overcrowding, unnatural circumstances (like the transfer of a troupe of rhesus monkeys to a new island), and deficient captive conditions play a role in most of these reports, with the latter blamed for incidents of infanticide in bush babies, lemurs, marmosets, and squirrel monkeys. In each case, the victims were invariably neonates, while the aggressors were either group members or relatives, including siblings. Responding to the problem, caretakers at facilities housing breeding primate colonies began isolating pregnant females before they gave birth, and these efforts have proven effective.

One primate group in which infanticide and cannibalism are relatively common practices is the chimpanzees (*Pan troglodytes*), and descriptions of the behavior among our closest relatives are both chilling and fascinating.

Initially, reports of chimpanzee cannibalism focused solely on adult males, who routinely killed and sometimes consumed infants belonging to "strangers" (i.e., adult females from outside their own groups). According to Dr. Jane Goodall, female chimpanzees sometimes transferred from one community to another. "A female who loses her infant during an encounter with neighboring males is likely to come into oestrus within a month or so and would then, theoretically, be available for recruitment into the community of the aggressors." Similar behavior is seen in bears and large cat species, like lions.

Other attacks by male chimps on infant-bearing females took place during "inter-community aggression" as occurs, for example, when groups of male chimpanzees patrolling the outer edges of their territories encountered individuals from adjacent communities.

Then, in 1976, Goodall reported on three observations in which two female chimps were involved in within-group infanticide and cannibalism in Tanzania's Gombe National Park. What made these

attacks unique was the absence of male involvement. Stranger yet was the fact that the individuals involved were a mother (Passion) and daughter (Pom), whose seemingly premeditated tag-team approach to somewhere between five and ten infant-bearing females provided researchers with a grim explanation for previously unexplained infant disappearances. Goodall believes that the attacks on the mothers functioned solely as a means to acquire food, since "once they had established their claim over their prey they made no further aggressive attacks on the mothers."

Thirty years later, similar attacks were carried out by female chimp coalitions against infant-bearing mothers in in Uganda's

Budongo Forest. A team led by comparative psychologist Simon Townsend believes that the lethal attacks were triggered by an influx of females, leading to increased competition for resources.

Although acts of cannibalism in chimpanzees are not everyday occurrences, some researchers have suggested that the encroachment of humans into the areas surrounding preserves inhabited by chimps will eventually lead to population density issues and more competition for dwindling resources. If this occurs, incidences of cannibalism by our closest relatives may be expected to increase.

# 5: Bear Down

*In Panama, I found a spider that eats it own limbs during lean times. I am told they grow back. But though the distinction is razor-thin, desperation is not the same thing as determination.*

—Taona Dumisani Chiveneko, *The Hangman's Replacement: Sprout of Disruption*, 2013

If you believe the news reports, this is not a good time to be a polar bear. Over the past several years, there have been dozens of headlines that ran something like this: "Polar Bears Are Turning to Cannibalism as Arctic Ice Disappears," "Is Global Warming Driving Polar Bears to Cannibalism?" "Polar Bear Cannibalism Linked to Climate Change."

As a vertebrate zoologist, I was interested in determining whether or not a transition in polar bear diets had actually taken place. And if it had, I wondered whether we were involved.

Polar bears (*Ursus maritimus*, Latin for "marine bear") are among the world's largest carnivores, a diverse mammalian order whose members include cats (felids), dogs (canids), raccoons (procyonids), and weasels and their relatives (mustelids). They are, of course, famous for their meat-eating diets and many of them share a characteristic known as *carnassial* teeth (or carnassials). In the majority of mammal species, when the jaw closes, the premolar and molar teeth on the upper jaw fit snugly into those on the lower jaw. This facilitates the crushing of food items before they're swallowed. In most carnivore species, though, when the jaws close, the last upper premolar and the first lower molar on each side shear past each other like blades, effectively slicing large pieces of meat into smaller pieces that can be readily swallowed. Carnassial dentition was lost in most bears as they evolved more omnivorous feeding habits. Here, the hard-to-digest plant material required a mash-up by more traditional molars, thus increasing its surface area and allowing for more efficient breakdown by enzymes like cellulase. In polar bears, however, fully functional carnassials have apparently re-evolved—a reflection of the species' strict meat-eating diet, which consists primarily of ringed seals (*Pusa hispida*) and bearded seals (*Erignathus barbatus*).

Secondarily evolved traits like carnassial teeth in polar bears are common in nature. For example, having inherited the ability to swim from their fishy ancestors, many ancient vertebrates lost the ability (and related features like fins) as they became more and more adapted to terrestrial lifestyles. Swimming re-evolved in some lineages, leading to creatures like seals and whales, whose fins are actually modified terrestrial limbs.

Cannibalism has been recorded in at least 14 species of carnivores. In pumas (*Puma concolor*), lynx (*Lynx lynx*), leopards (*Panthera pardus*), and sea lions (*Phocarctos hookeri*), it appears to occur for many of the usual reasons, including stress (due to lack of food), elimination of rivals, and increased mating opportunities.

Heterocannibalism, in this case, eating the cubs that another male sired, is clearly a reproductive strategy in male lions (*Panthera leo*) after taking over a pride. Through the practice of infanticide, the incoming males terminate the maternal investment in unrelated cubs. A lioness with cubs will not come into heat for a year and a half after giving birth, but similar to what has been observed in other mammals, a lioness that loses her cubs becomes sexually receptive almost immediately.

Interestingly, females of a similarly social predator, also found on the African savannah, possess a potent defense against infanticide and cannibalism. This adaptation has also enabled females of this species to become the dominant clan members. How this phenomenon works is fascinating, although it requires a brief review of development genetics. An additional bonus for this approximately 60-second commitment will be an answer to one life's great mysteries, namely, "Why do men have nipples (or penises, for that matter)?"

During early embryological development, mammal embryos are genderless. At a certain point, tiny buds of tissue grow into precursors of the penis and mammary glands. Sex determination is based on the embryo having one of two combinations of the X and Y sex chromosomes. These combinations, XY or XX, act like two versions of a blueprint. The XY blueprint results in the production of the male hormone testosterone, a chemical messenger that stimulates the growth of the penis bud into a penis. Since testosterone prevents the further development of the mammary glands, this explains why males still have the nipples they grew as genderless embryos—but don't produce milk. Alternately, having two X chromosomes results in the production of estrogens, the primary female sex hormones, and these stimulate the production of mammary glands. The female hormones also put a halt to the growth of the penis bud, leaving behind the clitoris, a tiny, erectile structure, which unlike the penis, is not involved in urination.

In the spotted hyena (*Crocuta crocuta*), the developmental scenario described above has been flipped on its ear. Scientists believe that at some point in their evolutionary past, a genetic mutation initiated the production of higher levels of the male sex hormone testosterone in *female* spotted hyenas. As a result, bulked up, hyperaggressive females dominate every interaction with their male counterparts, with males even coming up short in the external genitalia department. Female hyenas develop a remarkably elongated clitoris, which resembles a longer version of the penis. Additionally, the normally liplike vulva is fused closed, thus enabling females to urinate through their pseudopenises (or *pseudopenes*), the tips of which are also penetrated by the Real McCoy during copulation. Completing the he-man look, the sealed-up vulva forms a matching pseudoscrotum, within which deposits of fat stand in for a functioning pair of testicles.

The female hyena's uniquely shaped external sex organs actually gave rise to a myth that these mammals are hermaphrodites. Although this is definitely not the case, the birth process is an extremely painful and dangerous experience for first-timers, and by now you may have guessed the reason. Large, full-term hyena fetuses must pass through the clitoris, which, if things proceed smoothly, causes it to tear open. Reportedly, stillbirths and instances of maternal mortality during delivery are high, but after the successful birth of the first litter, the clitoris never fully closes again, making subsequent births somewhat easier. So, while some aspects of this adaptation sound counterproductive, the fact remains that *Crocuta crocuta* is the most successful mammalian predator in all of Africa. One reason may be related to the fact that, unlike in lions, there is little danger that males will attempt to kill and consume unrelated cubs. Females, on the other hand, have been known to do just that.

All right now, what about those polar bears?

In 2009, mainstream media outlets began reporting that polar bears were undergoing a serious change in dietary habits. The take-home message was that global warming had reduced the Arctic sea ice, thus resulting in shorter hunting seasons for the bears and fewer seal kills. As a consequence, the stressed-out bears were starving and resorting to cannibalism in order to survive. The problem with most of these stories was that the authors left out a rather important fact—and it was one that researchers have known for decades.

According to wildlife biologist Mitchell Taylor, "Polar bears will readily eat other polar bears when they can do so without excessive risk of injury." In fact, males of most North American bear species will kill and eat conspecific cubs pretty much whenever they can get their paws on them. Researchers believe that infanticide during the breeding season may provide males with "a reproductive opportunity as well as a nutritional reward" since like the previously described lionesses, female polar bears will come into estrus more quickly if their offspring have been killed. Because of this, cannibalism has been, and continues to be, one of the greatest contributors to bear cub mortality, especially just after leaving the maternity den. The threat from adult males is one of the key reasons that mother bears are so protective of their cubs and also explains why females give males such a wide berth when selecting maternity den sites.

Recently, another probable cause of polar bear cannibalism was added to the mix. Because of incomplete reporting by the media, though, and a tendency to stress sensationalism over detail, the result has been a cannibalism-themed fiasco.

The mess came about soon after the 2006 publication of a paper by Arctic researcher Stephen Amstrup. He and his coworkers were clearly alarmed by three incidents of cannibalism by polar bears in the southern Beaufort Sea, which occurred during a two-and-a-half-month period. Two of the incidents involved the death and partial consumption of adult female bears. In one, the female's body was found inside a maternity den that collapsed during an attack by a predatory male bear. In the second case, the female polar bear was killed on the sea ice, presumably not long after emerging from its den with a cub. In the third case, a one-year-old male was killed and partially consumed by an adult male. According to Amstrup and his colleagues, these attacks were unique because they had taken place in areas not generally frequented by male polar bears. Each year, once the Arctic sea ice melts and polar bears are forced onto the land, males are usually found near the coast while females and their cubs venture farther inland, and away from the males.

In the cases documented by Amstrup, the researchers concluded that "the underlying causes for our cannibalism observations are not known." They suggested that the incidents could have been "chance observations of previously unobserved rare events, or even a single rogue bear that adopted a [hunting] strategy including cannibalism."

What got the media machine cranking, though, was the researchers' hypothesis that these attacks and subsequent cannibalism might have resulted from male polar bears being "the first population segment to show adverse effects of the large ice retreats of recent years. . . . We hypothesize that nutritional stresses related to the longer ice-free seasons that have occurred in the Beaufort Sea in recent years may have led to the cannibalism incidents we observed in 2004."

The problem was not in the presentation of Amstrup's hypothesis, but the fact that many of the media reports that followed neglected to mention that cannibalism in polar bears was already known to be a naturally occurring event, with the first published report surfacing in 1897. By leaving out this vital fact, those working to publicize the effects of global climate change suddenly found themselves on the wrong end of some serious butt-kicking from climate change deniers. These zealots were quick to point out that cannibalism was quite common in polar bears and that the attempt to link polar bear cannibalism to what they referred to as the "Global Warming Hoax" was just another instance in which scientists were flat-out lying to the public. In reality, modern researchers have been reporting on non-climate-change-related infanticide and cannibalism in polar bears for decades, a point Amstrup and his coauthors also discussed their paper, and a point neglected in most of the media coverage.

Ultimately, though, the authors of the sensationalized headlines ignored that information. Instead they cobbled together their stories from non-scientific sources, including a short article by another non-scientist. This one warned of "GRAPHIC PHOTOS" and opened with the line, "Cannibalism is not part of the polar bears' M.O." As a result, a valid scientific hypothesis—*Global climate change has led to a reduction in Arctic sea ice, and this may be causing increased incidences of cannibalism in polar bears*—now takes a back seat to a distorted take on the subject as well as a deceptive but well executed argument by climate change deniers.

This would be my first experience with cannibalism-related sensationalism, but it would definitely not be the last.

# 6: Dinosaur Cannibals?

*Personally, I suspect that a whole pack of full-grown* T. rex *would have a very hard time finding enough to eat.*

—Paleontologist Nicholas Longrich, Discovery News, October 15, 2010

While we're on the topic of large, meat-eating animals embroiled in cannibalism-related controversies, I thought this would be the perfect time to bring up the topic of cannibalism in dinosaurs—or the lack thereof.

*Coelophysis bauri* was one of the earliest dinosaurs—a carnivorous and remarkably birdlike biped that lived approximately 200 million years ago across what is now the southwestern United States. A fast runner, it stood about a meter tall at the hips and had an overall length of about three meters from snout to tail. Equipped with a mouthful of recurved and bladelike teeth, *Coelophysis* was thought to feed on smaller animals like lizards.

In 1947, a team from the American Museum of Natural History (AMNH) working at Ghost Ranch in New Mexico unearthed a huge bone bed composed of hundreds of *Coelophysis* skeletons. After examining the fossils, famed AMNH paleontologist Edwin Colbert claimed that the abdominal cavities of some of the specimens contained the bones of smaller individuals of the same species.

Thus was born the "cannibal-*Coelophysis* hypothesis" and the subsequent portrayal of *Coelophysis* and other dinosaurs as cannibals. Reminiscent of the misconceptions concerning male-munching black widow spiders, the depiction of dinosaurs as cannibals remained unchallenged for decades.

In 2005, another group of researchers from the AMNH set out to determine whether or not claims of dinosaur cannibalism could be supported. Led by paleontologists Sterling Nesbitt and Mark Norell, they performed detailed morphological and histological analyses of the bones (something Colbert did not do). Soon enough, the scientists uncovered a slight problem—not only were the bones in question not from juvenile specimens of *Coelophysis*, they weren't even dinosaur bones. Instead, the fragments recovered from the abdominal cavities of the two relevant *Coelophysis* specimens belonged to crocodylomorphs, a group that includes crocodiles and their extinct relatives—but not dinosaurs.

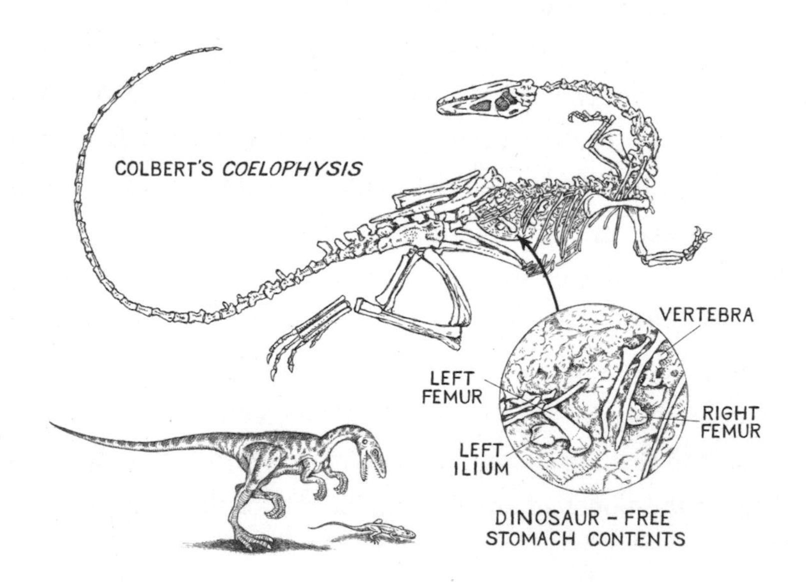

Investigating further, Nesbitt and Norell determined that another example of reputed dinosaur cannibalism was also problematic. In a much-publicized case, paleontologist Aase Jacobsen reported in 1998 that bite marks on the bones of tyrannosaurs from the Dinosaur Park Formation in Alberta, Canada, were consistent with bite marks from conspecifics. The AMNH researchers, however, called this cannibalism claim into question by pointing out that there were actually two tyrannosaur genera, *Gorgosaurus* and *Daspletosaurus*, living at the site. Since even by its loosest definition, cannibalism requires that the participants be conspecifics, the potential that the predation took place between separate species led Nesbitt and Norell to raise a warning flag. They concluded that while cannibalism "may be expected in non-avian dinosaurs," it was "not as prevalent as was once supposed."

In the limited realm of dinosaur cannibalism, an additional piece of evidence came from paleontologist Nicholas Longrich and several high profile co-authors in 2010. The researchers discovered bite marks on four museum specimens of *Tyrannosaurus rex*. They reasoned that since *T. rex* was the only large predator alive at that time, the score marks and gouges on the skeletons must have been made by conspecifics, and that the most likely scenario was that the marks had been made during scavenging of carcasses. Longrich and his coworkers concluded that their results provided solid evidence that "cannibalism seems to have been a surprisingly common behavior in *Tyrannosaurus*, and this behavior may have been relatively common in carnivorous dinosaurs."

I interviewed Mark Norell on a beautiful mid-September afternoon at the American Museum of Natural History. His lab is a dinosaur lover's dream—a remarkable, fossil-filled space that opens onto one of the museum's famous turrets. The view from Norell's high-ceilinged office was nothing short of spectacular, and as we talked it was difficult not to look over his shoulder at the wide swath of Central Park below us.

"I think there's very little evidence at all for dinosaur cannibalism," Norell told me. "Although a lot of it really depends on what you'd call cannibalism. If a tyrannosaur dies and another tyrannosaur comes along and eats it, is that cannibalism? Or is that just scavenging a dead carcass? I have a picture around here someplace of a camel eating a dead camel that was lying there. Is that cannibalism?"

I told him that I didn't consider hunting and killing to be prerequisites for cannibalism and that, from what I'd learned, scavenging your own species *was* cannibalism. I used the example of besieged cities, where the victims of starvation or exposure were consumed, sometimes by their own relatives.

As scientists began to study cannibalism in nature, some of the behaviors they observed expanded on the definition of cannibalism coined by Elgar and Crespi in the early 1990s. The requirement that the cannibal kill the conspecific before eating it was dropped (enabling certain forms of scavenging to be considered cannibalism), as was the necessity that the cannibal consume the entire victim. As a result, some forms of behavior now straddle a line between cannibalism and something else—in this case, scavenging.

But even allowing for a broad definition of cannibalism, in the case of dinosaurs, Norell had some major problems with the evidence that Longrich had presented. For example, he said that the gouges on Longrich's *T. rex* bones could have been inflicted by conspecifics fighting each other, not necessarily eating each other.

According to Norell, the only compelling evidence for dinosaur cannibalism appeared to have occurred in the late Cretaceous theropod *Majungasaurus crenatissimus*, an example uncovered by geologist Raymond Rogers from multiple bone beds from a Madagascan rock formation thought to be between 70.6 and 65.5 million years of age.

In a 2014 phone conversation, I asked Rogers how he had come to the conclusion that *Majungasaurus* was a cannibal. He explained that he'd been looking at dinosaur bones for decades and that specimens from a particular region in Madagascar (dated to approximately 70 million years ago) had a remarkably high number of teeth marks.

"Quickly enough, you start thinking about who could have made these bite marks, and [in this instance] there are very few candidates." According to Rogers, there were only a few large carnivores living at the site, which they named MAD05-42. One was a crocodile, "which would have made no comparable traces on the bones" and the other a small theropod dinosaur, "which had tiny little teeth."

"And then you've got *Majungasaurus*, which had large teeth. When you look at the scoring patterns on [*Majungasaurus*] bones you can match them to the spacing of *Majungasaurus* teeth and their actual denticle patterns."

"So there's no potential that you might be missing another large predator—something you just haven't dug up yet?" I asked.

"Right, I don't think we're missing anything. And if we are missing something it would have to be big, and arguably it would have to be rare. But the bite marks are anything but rare. So . . . whatever it is, it would have to be really big, really cryptic, really rare, and it would have to bite everything, which doesn't make any sense. Basically, there are tons of bite marks that match the teeth of *Majungasaurus* and nothing else matches those traces."

I pressed on. "So how do you know that these tooth marks on *Majungasaurus* bones weren't made during combat between conspecifics?"

Rogers said he had good evidence to prove that hadn't been the case. Elements of the vertebral column showed evidence of having been scraped and possibly gnawed. This type of "late stage scavenging" took place after the limb muscles and the guts had been consumed, and when the scavenger had to work hard to obtain any further nutrition from the carcass.

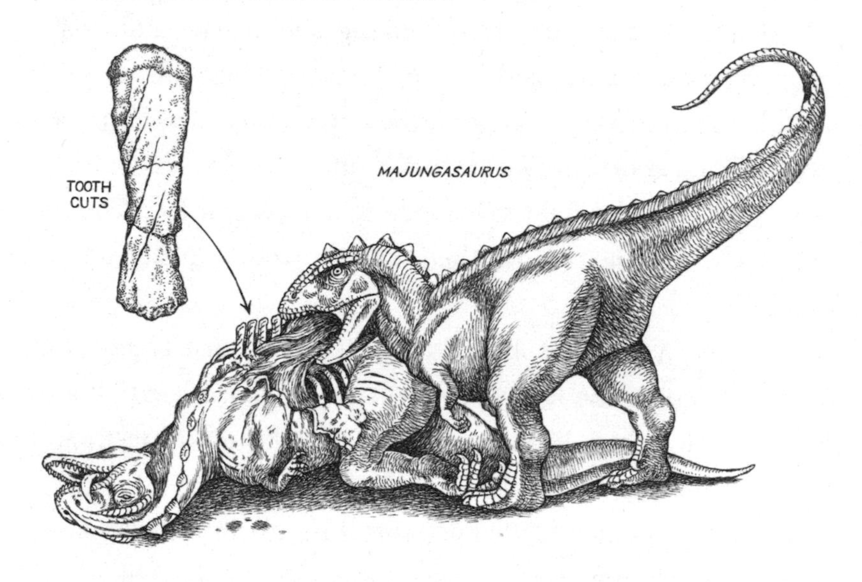


"There's no evidence, whatsoever, for killing or jostling with conspecifics. There's evidence for feeding. And the evidence we have for feeding is consistent with some pretty focused effort. These bones were scraped."

It was beginning to sound like *Majungasaurus* had been following two of Gary Polis's cannibalism-related generalizations: first, that incidents of the behavior increased during stressful environmental conditions; and second, that the absence of alternative food sources often led to cannibalism. So I asked Rogers if he believed that stressful environmental conditions could have driven *Majungasaurus* to scavenge conspecifics.

"It kind of looks that way," he replied. "The overall reconstruction that we put together is pretty well supported. It seems like these ancient ecological systems were devastated again and again and a lot of things died. I think that if you're a large meat-eating dinosaur, you'll capitalize on whatever there is to feed upon, and I have no doubts that a creature like *Majungasaurus* would take the resources that were available."

Finally, I asked the paleontologist how prevalent he thought cannibalism had been among the dinosaurs.

"I doubt that it was uncommon," Rogers replied. "What's uncommon is evidence. Because I think that more often than not, animals can probably get their meals without heavily working bone, and when you're dealing with fossils . . . that's all you got. If it's not written on the bone, you're not going to get the story. So . . . I don't think cannibalism was unique to any particular theropod group. When you drive around Montana today and see ground squirrels eating ground squirrels and you read papers about dogs eating dogs, and lizards eating lizards . . . it's pervasive. I have no doubts that cannibalism was widely practiced by dinosaurs. The fact that there have only been two cases of dinosaur cannibalism . . . that's just an artifact of paleontology and the [scarcity of the] fossil record."

So was dinosaur cannibalism a rare event? There appear to be two contrasting issues here. First, knowing what we do about the prevalence of cannibalism in the animal kingdom, it makes sense that dinosaurs might have exhibited the behavior for the same reasons other animals do—namely overcrowding, predation, competition, and hunger. Alternately, cannibalism is relatively rare in birds, the only surviving link to the Mesozoic dinosaurs. The problem with comparing the two groups, though, is that birds are gape-limited predators while carnivorous dinosaurs certainly were not.

Then there's the lack of widespread physical evidence that dinosaur cannibalism occurred, and because of this, some paleontologists, like Mark Norell and his colleagues, are unwilling to make any sort of speculative leap. Instead their explanations for the strange bite marks on ancient *T. rex* bones fall back onto behavior they do have evidence for—like fighting. According to Norell, "There's nothing like a smoking gun that anyone has ever presented to me and said, 'This is it!'"

"Can you give me an example of a paleontological 'this is it' moment?" I asked.

Norell brought up the now-generally-accepted claim that modern birds were actually theropod dinosaurs and would, therefore, show similar anatomical and behavior traits.

"We found evidence that dinosaurs sat on their nests," he said. "We were able to show that. Dinosaur feathers—we were able to show that. In our lab we tend to be incredibly careful about what we say about this stuff . . . but no one has ever been able to come up with a *total* case for dinosaur cannibalism, like a member of the same species that's inside the body cavity. Like *Coelophysis* was *supposed* to be."

In a telling side story, Raymond Rogers provided another example of just how attractive the word *cannibalism* is to the media. "I took this story of dinosaur cannibalism to the Society of Vertebrate Paleontology meetings, and I called it "Conspecific Scavenging," which is what I think it is. I remember that a guy from *Science News* looked at it, but nobody else really took much notice at all. I went home and thought about it, and I was like, 'You know, why don't I just call it cannibalism?' So I did . . . and after that the story got in *Nature* and it was on the front page of Google News for about a week."

"Well there you go," I responded with a laugh. "That word does set something off in us."

"Right," Rogers agreed. "Before I knew it, *USA Today* was talking about dinosaurs, chianti, and fava beans."

# 7: File Under: Weird

*Cannibalism is found in over 1,500 species. Anthropophagusphobia (fear of cannibals) is found in only one. Which seems unnatural now?*

—Author unknown

Is eating one's own fingernails or mucus an example of auto-cannibalism? And what about breast-feeding? Is this type of parental care actually a form of cannibalism? Raymond Rogers considers scavenging the body of a conspecific dinosaur a form of cannibalism. Mark Norell, not so much. All are examples of a gray area between what most people consider cannibalism and other forms of behavior.

Like breast-feeding, the following example is a form of parental care, but one that extends further into the realm of cannibalism-related behavior. It occurs in the caecilians, a small order of not-very-obvious amphibians, whose legless bodies often get them mistaken for worms or snakes. Caecilians inhabit tropical regions of Central and South America, Africa, and Southern Asia—a neat trick that definitely lends support to the theory of continental drift. Although some caecilians are aquatic, it is not believed that their ancestors were strong enough swimmers to cross the Atlantic Ocean. Instead, prehistoric caecilians were likely separated when the current continents of South America and Africa split apart between 100 and 130 million years ago.

Caecilians also serve as great examples of convergent evolution, in which unrelated organisms each evolve similar anatomical, physiological, or behavioral characteristics, because they inhabit similar environments. As a result of their subterranean lifestyles, caecilians share a number of anatomical similarities with moles and mole rats. In each, the eyes are either set deeply into the skulls or are covered by a thick layer of skin, and as a consequence they are nearly blind.

Caecilians also possess a pair of short "tentacles" located between their nostrils and eyes. These chemosensors enable the subterraneans to "taste" their environments without opening their mouths, as they burrow through the soil or leaf litter in search of insects and small vertebrates. Similar types of sensory structures can be seen in other burrowing creatures, most notably the aptly named star-nosed mole (*Condylura cristata*).

As a group, caecilians exhibit a fair degree of reproductive diversity (which will become an important aspect of their cannibal-related behavior). Approximately half of the 170 species are *oviparous* (egg layers), and hatchlings either resemble miniature versions of their parents or pass through a brief larval stage. Other species are *viviparous*, giving birth to tiny, helpless young.

All caecilians do share one characteristic unique to the amphibians: internal fertilization, and during this process, sperm is deposited into the female's cloaca with the aid of a penis-like structure called a phallodeum. For the orifice-challenged, a reminder that in many vertebrates (like amphibians, birds, and reptiles), the cloaca is a single opening shared by the intestinal, reproductive, and urinary tracts.

But as interesting as the concept of legless caecilians wielding their penises underground might be (admittedly, it disturbed some of my older Italian relatives until I explained the spelling differences), information about caecilian cannibalism began emerging from Marvalee Wake's lab at the University of California, Berkeley. The herpetologist extraordinaire was looking at fetal and newborn individuals from several viviparous species and began investigating the function of their peculiar-looking baby teeth (better known to scientist-types as deciduous dentition).

While some of the teeth were spoon-shaped, others were pronged or resembled grappling hooks, but none of them resembled adult teeth. Wake also performed a microscopic comparison of caecilian oviducts. She observed that in pregnant individuals, the inner (i.e., epithelial) lining of the oviduct was thicker and had a proliferation of glands, which she referred to as "secretory beds." These glands released a substance that fellow researcher H.W. Parker had previously labeled "uterine milk." Parker described the goo, which he believed the fetuses were ingesting, as "a thick white creamy material, consisting mainly of an emulsion of fat droplets, together with disorganized cellular material." He also thought that the caecilians' fetal teeth were only used *after* birth, as a way to scrape algae from rocks and leaves. Wake, however, had her doubts, especially since she noticed that these teeth were resorbed before birth or shortly after.

Pressing on with her study, Wake saw something odd. In sections of oviduct adjacent to early-term fetuses, the epithelial lining was intact and crowded with glands. However, in females carrying late-term fetuses, the lining of the oviduct was completely missing in the areas adjacent to the fetuses, although it was intact in regions well away from the action. Wake proposed that fetal caecilians used their teeth before birth to scrape fat-rich secretions and cellular material from the lining of their mother's oviduct. Although this behavior couldn't be seen directly, she had gathered circumstantial evidence in the form of differences in the oviduct between early-term and late-term individuals. After an analysis of fetal stomach contents revealed cellular material, Wake had enough evidence to conclude that caecilian parental care extended beyond the production of uterine milk and into the realm of cannibalism. Unborn caecilians were eating the lining of their mothers' reproductive tracts.

But if the consumption of maternal epithelial cells in viviparous caecilians gave this admittedly strange behavior a cannibalistic slant, it was in the egg-laying species that the story really took off.

In 2006, caecilian experts Alexander Kupfer, Mark Wilkinson, and their coworkers were studying the oviparous African caecilian, *Boulengerula taitanus*, when they made a remarkable discovery. This species had been previously reported to guard its young after hatching, and the researchers wanted to examine this behavior in greater detail. They collected 21 females and their hatchlings and set them up in small plastic boxes designed to resemble the nests they had observed in the field. Their initial observations included the fact that the mothers' skin was much paler than it was in non-moms and that hatchlings had a full set of deciduous teeth resembling those employed by their oviduct-munching cousins.

Intrigued, the researchers set out to film the parental care that had been briefly described by previous workers. On multiple occasions, Kupfer and Wilkinson observed a female sitting motionless while the newly hatched brood (consisting of between two and

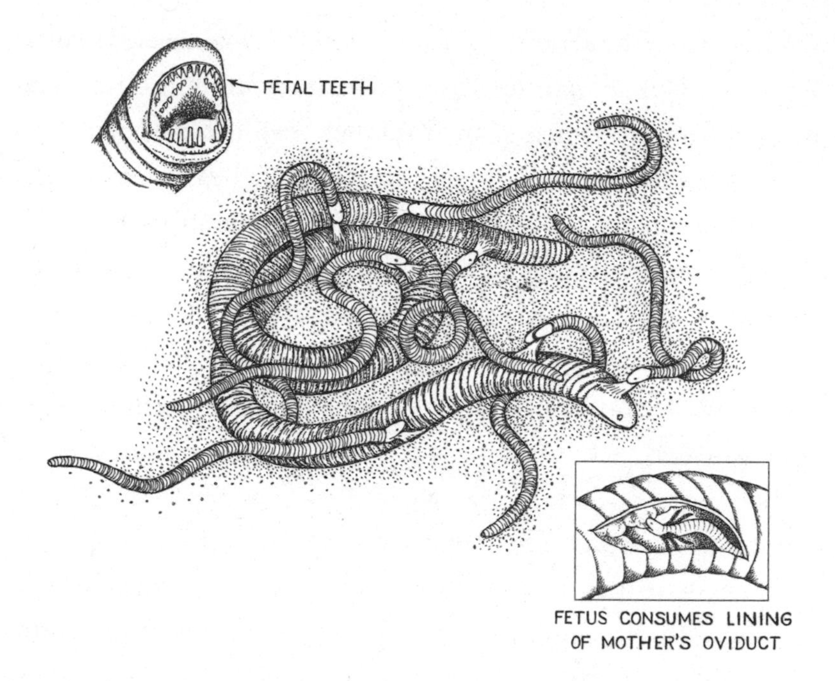


nine young) slithered energetically over her body. Looking closer, they noticed that the babies were pressing their heads against the female, then pulling away with her skin clamped tightly between their jaws. As the researchers watched, the baby caecilians peeled the outer layer of their mother's skin like a grape . . . and then they consumed it.

Scientists now know that these bouts of "dermatophagy" reoccur on a regular basis and that the mothers' epidermis serves as the young caecilians' sole source of nutrition for up to several weeks. For their part, female caecilians are able to endure multiple peelings because their skin grows back at a rapid rate.

"The outer layer is what they eat," Wilkinson said. "When that's peeled off, the layer below matures into the next meal."

In addition to the ability of the skin to quickly repair and replenish itself, the nutritional content of this material is yet another interesting feature in this bizarre form of parental care. Normally, the outermost epidermal layer, the stratum corneum, is composed of flattened and dead cells whose primary functions are protection and waterproofing. But when the researchers examined the skin of brooding female caecilians under the microscope, they noticed that the stratum corneum had undergone significant modification. Not only was the layer thicker, it was also heavily laden with fat-producing cells, which explained why the baby caecilians experienced significant increases in body length and mass during the weeklong observations. It also explained why mothers of newly hatched broods experienced a concurrent decrease in body mass of 14 percent. In short, dermatophagy is a great way to fatten up the kids, but for moms on the receiving end of their gruesome attentions, the price is steep.

Scientists now believe that the presence of dermatophagy in both South American and African oviparous species offers strong support for the hypothesis that these odd forms of maternal investment originally evolved in the egg-laying ancestor of all modern caecilian species. Consequently, when the first live-bearing caecilians evolved, their unborn young were already equipped with a set of fetal teeth, which took on a new function, allowing them to tear away and consume the lining of their mothers' oviduct.

# 8: Neanderthals and the Guys in the Other Valley

*Here is a pile of bones of primeval man and beast all mixed together, with no more damning evidence that the man ate the bears than that the bears ate the man—yet paleontology holds a coroner's inquest in the fifth geologic period on an 'unpleasantness' which transpired in the quaternary, and calmly lays it on the MAN, and then adds to it what purports to be evidence of CANNIBALISM. I ask the candid reader, Does not this look like taking advantage of a gentleman who has been dead two million years.*

—Mark Twain, *Life As I Find It*, 1871

In 1856, three years before publication of Charles Darwin's *On the Origin of Species*, a worker at a limestone quarry near Düsseldorf, Germany, uncovered the bones of what he thought was a bear. He gave the fossils to an amateur paleontologist, who in turn showed them to Dr. Hermann Schaaffhausen, an anatomy professor at the University of Bonn. The bones included fragments from a pelvis as well as arm and leg bones. There was also a skullcap—the section of the cranium above the bridge of the nose. The anatomist immediately knew that while the bones were thick and strongly built, they had belonged to a human and not a bear. They were, though, unlike any human bones he had ever seen. Beyond the robust nature of the limbs and pelvis, the skullcap had a low, receding forehead and a prominent ridge running across the brow. These anatomical differences led him to conclude that these were the remains of a "primitive" human, "one of the wild races of Northern Europe."

The next year, the men announced their discovery in a joint paper, but the excitement they hoped to generate never materialized. This was, after all, a scientific community that had yet to reject the concept that organisms had not changed since God created them only five thousand years earlier. It was no real surprise, then, when a leading pathologist of the day examined the bone fragments and pronounced them to be modern in origin, insisting that the differences in skeletal anatomy were pathological in nature, having been caused by rickets, a childhood bone disease. He blamed the specimen's sloping forehead on a series of heavy blows to the head.

By the early 1860s, thanks to the publication of *On the Origin of Species*, there was increased interest in evolution, especially the topic of human origins. Now the concept of "change over time" was no longer alien, and in the newly minted Age of Industry, the idea of the survival of the fittest was not only palatable, it was profitable. By 1864, the rickets/head injury hypothesis had been overshadowed by the discovery of new specimens with identical differences in skeletal structure. Neanderthal Man became the first prehistoric human to be given its own name, a moniker derived from the Neander River Valley, where the presumed first fossils had been uncovered.[11]

Thrust into the scientific and public eye, Neanderthal Man became a Victorian era sensation. Scientists like Darwin's contemporary and friend Thomas Huxley believed these particular remains were important because they established a fossil record for humans that supported Darwin's newly published theory. With none of his friend's famous restraint, Huxley announced that *Homo sapiens* had descended with modification from apelike ancestors, and the Neanderthals were just the proof he needed.

Huxley's rationale was that, although Neanderthals shared many characteristics with modern humans, they also exhibited primitive traits, thus serving as physical evidence that humans, like other organisms, had evolved gradually and over a vast time frame. Neanderthals, he reasoned, were a part of Darwin's branching evolutionary tree, with this particular branch leading to modern humans.

The most serious argument against Huxley's hypothesis was put forth in 1911. Marcellin Pierre Boule, a French anthropologist and scientific heavyweight, had been called upon to study and reconstruct a Neanderthal specimen that had been uncovered in France several years earlier. Once Boule was finished, anyone viewing the reconstruction would come away with some strong ideas about

---

11 In the early 20th century, "Thal," the German word for "valley" was changed to "Tal." As a consequence, "Neandertal" is a common alternative to "Neanderthal." Since the scientific name for the species (or subspecies) remained *neanderthalensis*, most scientists do not use the new spelling. Soon after the name was coined, researchers determined that two other collections of strange bones found decades earlier in Belgium and Gibraltar (and unnamed by those who discovered them), were also the remains of Neanderthals.

what Neanderthals looked like. Significantly, he gave the skeleton a curved rather than upright spine, indicative of a stooped, slouching stance. With bent knees, flexed hips, and a head that jutted forward, Boule's Neanderthal resembled an ape. The anthropologist also claimed that the creature possessed the intelligence (or lack of intelligence) to match its apelike body.

Boule commissioned an artist to produce an illustration of his reconstruction, and the result depicted a hairy, gorillalike figure with a club in one hand and a boulder in the other. The creature stood in front of a nest of vegetation, another obvious reference to gorillas. Boule's vision of Neanderthals, with their knuckle-dragging posture and apelike behavior, also left an indelible mark on a public eager to hear about its ancient ancestors. For decades to come, Neanderthals would become poster boys for stupidity and bad behavior. The epitome of a shambling, dimwitted brute, "Neanderthal" became synonymous with "bestial," "brutal," savage," and "animal."[12] In "The Grisly Folk," an influential story written by H. G. Wells in 1921, the author stuck to the Boule party line, depicting "Neandertalers" as cannibalistic ogres: "when his sons grew big enough to annoy him, the grisly man killed them or drove them off. If he killed them he may have eaten them." According to Wells, the grisly men also developed a taste for the modern humans who had moved into the neighborhood, finding "the little children of men fair game and pleasant eating." Because of this type of rude behavior ("lurking" was also a popular activity), Wells felt that the

12 Today, even among scientists and academics, calling someone a Neanderthal rarely implies that we're referring to a skilled hunter who uses his oversized brain to fashion and employ an array of sophisticated tools.

ultimate extermination of the Neanderthals was completely justified, allowing modern humans to rightfully inherit the Earth.

The only problem with Wells's character, according to paleontologist Niles Eldridge, was that it was based on Boule's misconceptions. "Every feature that Boule stressed in his analysis can be shown to have no basis in fact."

Since the early 20th century, Neanderthals have undergone a further series of transformations and today there are two main hypotheses.

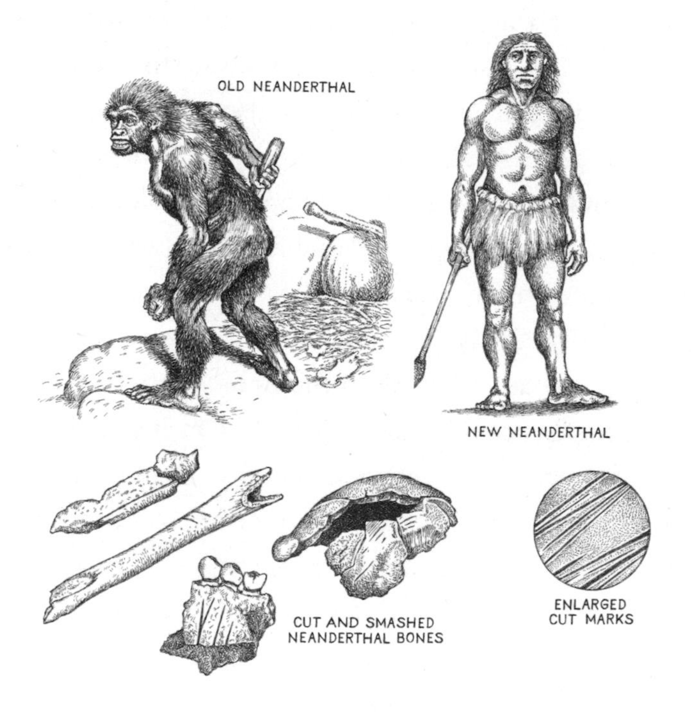


That they were our direct ancestors (*Homo sapiens neanderthalensis*) became part of what is known as the Regional Continuity hypothesis. It is a view currently supported by paleoanthropologist Milford Wolpoff, who believes that Neanderthals living in

Europe and the Middle East interbred with other archaic humans, eventually evolving into *Homo sapiens*. According to Wolpoff, similar regional episodes took place elsewhere around the globe as other archaic populations intermingled, hybridizing into regional varieties and even subspecies of humans. Importantly, though, there would be enough intermittent contact between these groups (Asians and Europeans, for example) so that only a single species of humans existed at any given time.

Alternatively, the Out of Africa hypothesis holds that modern humans evolved once, in Africa, before spreading to the rest of the world where they displaced, rather than interbred with, the Neanderthals (*Homo neanderthalensis*) and others who had been there previously. The groups driven to extinction by *Homo sapiens* had themselves evolved from an as-yet-undiscovered species of *Homo* (perhaps *H. erectus*) that had originated in Africa and migrated out earlier.

I interviewed Ian Tattersall in his impressively cluttered office at the American Museum of Natural History, where he is curator emeritus in the Division of Anthropology. Tattersall is one of the world's eminent paleoanthropologists and has authored (or coauthored) more than a hundred articles on ancient hominids, as well as popular books on the topic. He is a major proponent of the Out of Africa hypothesis.

As we sat surrounded by fossilized bones, ancient artifacts, and a small population of lifelike hominid replicas, I asked Tattersall if he thought that Neanderthals and *Homo sapiens* had interbred and hybridized, the central tenet of the Multiregional (or Regional Continuity) hypothesis.

"Not to any biologically significant extent, no," he replied. "Neanderthals were incredibly different, and I think they viewed the world very differently, too. Now there may have been a bit of Pleistocene hanky-panky that went on, but I don't think it was anything that would have affected the future trajectory of either population."

"So you don't subscribe to the 'Neanderthals-R-Us' hypothesis?"

"Look, you're a zoologist," Tattersall replied. "You're not tainted by the perceived wisdom of certain paleoanthropologists. Just have a look at these guys and make your mind up for yourself. Structurally, anatomically, and presumably behaviorally, too, Neanderthals and modern humans were very, very different. If you didn't have all these preconceptions we have from paleontological tradition, if you were comparing a Neanderthal and a modern human, you'd probably put them in different genera. They're *very* distinctive. They looked different and they behaved different. So they may have interacted to some extent, and they may have even exchanged the odd gene or two, but they certainly didn't just blend into each other. I just don't think that that's plausible at all."

As evidence, Tattersall cites the fact that in many places in the Middle East and Mediterranean Basin, Neanderthals and modern humans coexisted as morphologically distinct populations "for something like 50,000 years . . . so the idea of a gigantic, late-Pleistocene love-in among morphologically differentiated hominids simply defies every criterion of plausibility."

Supporting this stance are recent morphological and mitochondrial DNA studies that indicate a clear distinction between modern humans and Neanderthals. Further support came from Swiss researchers Mathias Currat and Laurent Excoffier. In 2011, they used admixture/competition models to calculate the amount of inbreeding that would have occurred when previously separated populations of Neanderthals and Paleolithic humans begin encountering each other. Modeling results showed that low levels of Neanderthal ancestry in Eurasians (2 to 3 percent of the genome) were compatible with an extremely low incidence of interbreeding, with successful couplings between Neanderthals and Paleolithic humans estimated to take place only once every 23 to 50 years.

But whatever hypothesis anthropologists choose to support concerning interactions between Neanderthals and modern humans, and the ultimate demise of the former, Neanderthals are no longer depicted as knuckle-dragging brutes. Instead, studies have shown that they were highly intelligent, with some specimens exhibiting a cranial capacity (i.e., brain volume) 100 to 150 milliliters greater than the 1,500-milliliter capacity of modern humans! Researchers have also learned that Neanderthals used fire, wore clothing, and constructed an array of stone tools, including knives, spearheads, and hand axes.

The possibility that Neanderthals practiced cannibalism was briefly argued in 1866 and again in the 1920s, after a fossil skull discovered in Italy was observed to have a gaping hole above and behind the right eye. The wound was initially interpreted as evidence that the skull had been broken open by another Neanderthal intent on extracting the brain for food, but researchers now believe that a hyena caused the damage.

More recent and significantly stronger evidence for Neanderthal cannibalism came from multiple sites in northern Spain, southeastern France, and Croatia. In each instance, Neanderthal bones exhibited at least some of the characteristics interpreted by anthropologists as "patterns of processing." This term refers to the telltale damage found on the bones of animals that have been consumed by humans. This Neanderthal-inflicted damage includes some combination of cut marks, which result when a blade is used to remove edible tissue like muscle; signs of gnawing or peeling; percussion hammering (abrasions or pits that result from the bone being hammered against some form of anvil); burning; and the fracturing of long bones, presumably to access the nutrient-rich marrow cavity.

But even when these patterns of processing are observed, researchers must proceed with caution before making claims about the occurrence of cannibalism. While these forms of bone damage can be strong indicators of human activity, they can also result from human behavior or phenomena completely unrelated to cannibalism. According to anthropologist Tim White, "Bodies may be buried, burned, placed on scaffolding, set adrift, put in tree trunks or fed to scavengers. Bones may be disinterred, washed, painted, buried in bundles or scattered on stones." In what is known as secondary burial, bodies that have already been buried or left to decompose are disinterred and subjected to additional handling. For the ancient Jews this involved placing the bones into stone boxes called ossuaries. For some Australian aboriginal groups and perhaps the ancient Minoans, secondary burial practices included the removal of flesh and cutting of bones. Rituals like these make it extremely difficult to distinguish between funerary rites and cannibalism, especially if the group in question no longer exists or if the rites are no longer practiced.

Cut marks on bones may be the result of violent acts related to war or murder. If you can imagine someone unearthing the skeleton of a soldier killed by a bayonet or sword, they might misinterpret the cut marks on the bones as evidence of cannibalism.

Clearly, then, blade marks and other damage inflicted on Neanderthals by conspecifics and other ancient human groups may have been caused by a variety of actions. Archaeologists now consider this type of osteological damage to be strong evidence for cannibalistic behavior only when it can be matched to similar damage found on the bones of game animals uncovered at the same site. The implication is that if animal and human bodies were processed in the same manner, and if the remains were discarded together, it is reasonably certain that cannibalism took place.

This appears to have been precisely what happened at a Neanderthal cave site known as Moula-Guercy in southeastern France. An excavation begun there in 1991 revealed the remains of six Neanderthals and at least five red deer (*Cervus elaphus*) dating to approximately 100,000 years before the present. The bones were distributed together and butchered in a similar fashion. The long bones and skulls were smashed, and telltale cut marks on the sides of the skulls indicated that the large jaw closure muscles had been filleted. There were also characteristic patterns of modification on the lower jaws, providing evidence that the tongues had been removed. Both Neanderthal and deer bones also exhibited peeling and percussion pits. Lastly, there were distinctive patterns of cuts indicating that bodies from both species had been disarticulated at the shoulder, a process that would have made carrying and handling them easier. According to Tim White, "The circumstantial forensic evidence [for cannibalism at Moula-Guercy] is excellent."

Of course there is always the possibility that this type of damage to the animal bones took place during butchery, but that the same types of stone tools were also used to deflesh and disarticulate human remains during non-cannibalistic mortuary practices. Echoing what Mark Norell would have considered "the smoking gun" for dinosaur cannibalism, anthropology writer Paul Bahn wrote that, "The only definitive evidence for prehistoric cannibalism would be the discovery of human remains inside fossil faeces or inside a human stomach."

But among anthropologists, even this type of evidence sparked a controversy.

In 2000, researchers working in the Four Corners region of the American Southwest reported that human myoglobin (a form of hemoglobin found in muscles) had been identified from a single fossilized coprolite described as being "consistent with human origin." The petrified poop had evidently been deposited onto a cooking hearth belonging to prehistoric Puebloans (Anasazi) sometime around 1150 CE. Together with defleshed human bones and butchering tools coated with human blood residue, the 30-gram fecal fossil was used to support the claim that cannibalism had taken place at the southwestern Colorado site known as Cowboy Wash. It is a finding that has been the subject of considerable debate, with some researchers insisting that the bone and blood evidence could also have resulted from corpse mutilation, ritualized executions, or funerary practices.

These scientists also point out that while the myoglobin in the coprolite was certainly human in origin, the animal that produced the feces was never positively identified. This raises the possibility that a coyote or wolf consumed part of a corpse and subsequently defecated in the abandoned cooking hearth.

Even with a set of paleoanthropological safeguards in place, mistakes can still occur. Some of these have been the result of bad writing, while in other instances, further research led to the discovery of additional non-cannibalism-related alternatives, these having nothing to do with animal or human interactions.

"In many cases you're finding bones in the normal paleontological environment," Ian Tattersall explained. "That is to say, they've all been scattered and they've been concentrated by water or whatever's happened to them, which had nothing to do with the actual human activities that may or may not have been carried out after they were deceased."

To envision how this "scattering" or concentration of fossils can occur, picture a stream cutting through a fossil-containing layer of rock. As the stream walls gradually wear away, fossils are exposed, washed out, and deposited into the stream bed randomly and over time. Similarly, different parts from the same organism might be exposed at different times, which can also lead to fragments from a single individual being scattered across a wide area.

This water-assisted movement can also take place *before* the specimens are fossilized. For example, the bodies of creatures that died along the length of an ancient body of water—or in it—may have been carried away by currents and deposited together due to gravity or the physical properties of the stream or river. If sediments covered the bodies rapidly enough, they may have become fossilized, but their final location may have little or nothing to do with the behavior and associations that took place when the organisms were alive. For this reason, archaeologists must be cautious when animal and human bones are found mixed together. The mélange does not necessarily prove that humans did the mixing.[13]

One instance in which the evidence for human cannibalism

---

13 Similarly, cave collapses also appear to have caused the demise of a number of Neanderthals whose fossilized bones were initially thought to exhibit percussion pits.

remains solid involves *Homo antecessor* ("pioneering man"), the reputed ancestor of Neanderthals. The first fossils of this species were uncovered in the 1980s in Atapuerca, a region in northern Spain. Initially, spelunkers found the bones of extinct cave bears at the bottom of a narrow 50-foot-deep pit. Excavation of the pit, now known as Sima de los Huesos (Pit of the Bones), was initiated in 1984 by paleontologist Emiliano Aguirre. After his retirement, Aguirre's students continued to work at the site, and in 1991 they began emerging from the stifling heat and claustrophobic conditions with well-preserved hominid bones.

Since then, the site has yielded more than 5,000 bone fragments from approximately 30 humans of varying age and sex. The researchers noted that some aspects of the skull and post-cranial skeleton appeared to be Neanderthal-like (including a large pelvis that someone christened "Elvis"). Eventually, though, the remains from Atapuerca exhibited sufficient anatomical differences from Neanderthals to warrant placing them into a separate species.

According to Ian Tattersall, *Homo antecessor* was "almost Neanderthal but not quite. . . . These guys were on the way to becoming Neanderthals." To the surprise of researchers, the remains of *Homo antecessor* recovered from Sima de los Huesos were dated to a minimum of 530,000 years, indicating that the Neanderthal lineage had been in Europe 300,000 to 400,000 years before the first Neanderthals, far longer than anyone imagined.

By 1994, researchers were claiming that *Homo antecessor* remains showed evidence of having been cannibalized. In this case, the fracture patterns, cut marks, and "tool-induced surface modification" were identical to the damage found on the bones of non-human animals that had presumably been used as food. All of the bones (human and non-human) were randomly dispersed as well. The researchers at Atapuerca concluded that the *H. antecessor* remains came from "the victims of other humans who brought bodies to the site, ate their flesh, broke their bones, and extracted the marrow, in the same way they were feeding on the [animals] also preserved in the stratum."

Interestingly, the presence of so many types of game animals led the same researchers to suggest that Atapuerca did not represent an example of stress-related survival cannibalism, and Tattersall agreed. "Sometimes the environment was pretty rich and you wouldn't necessarily need to practice cannibalism to make your metabolic ends meet, as it were. You'd be able to relatively easily find sources of protein otherwise."

Accordingly, the Neanderthal ancestors living at Atapuerca were likely not prehistoric versions of the Donner Party—stranded in horrible conditions and compelled by starvation to consume their dead. Instead, *Homo antecessor*, like many species throughout the animal kingdom, may have simply considered others of their kind to be food. In other words, they may have eaten human flesh because it was readily available and because they liked it.

No one is absolutely certain when the transition from *Homo antecessor* to Neanderthal Man took place, but it probably happened sometime around 150,000 years ago. If one does not subscribe to the idea that Neanderthal genes were eventually overwhelmed through interbreeding with their more intelligent cousins, then *Homo neanderthalensis* appears to have gone extinct approximately 30,000 years ago. I asked Tattersall to elaborate.

"Neanderthals and modern humans [i.e., *Homo sapiens*] managed to somehow partition the Near East among themselves for a long, long period of time, at a time when modern humans were not behaving like they do today."

"How did these humans differ from us?" I asked him.

"They left no symbolic record [e.g., depictions of their behavior and beliefs]. As soon as they started leaving a symbolic record, the Neanderthals were out of there."

I told Tattersall that I still didn't see the connection, so he explained it further. "I think that by the time the Neanderthals' homeland in Europe was invaded by modern humans, humans were behaving in the modern way and had become insuperable competitors."

Given what we know about modern humans and their treatment of the indigenous groups they encountered, it's difficult to argue against Tattersall's conclusions. In all likelihood, the Neanderthal homeland was indeed invaded by an advanced, symbolism-driven species and, as we'll see in the following chapter, it would have been more of a surprise in fact if *Homo sapiens* hadn't raped, enslaved, and slaughtered the Neanderthals and other groups they encountered there.

# 9: Columbus, Caribs, and Cannibalism

*The captain . . . took two parrots, very large and very different from those seen before. He found much cotton, spun and ready for spinning; and articles of food; and he brought away a little of everything; especially he brought away four or five bones of the arms and legs of men. When we saw this, we suspected that the islands were those islands of the Caribe, which are inhabited by people who eat human flesh.*

—Dr. Diego Álvarez Chanca

The quote above came from Dr. Diego Álvarez Chanca, a physician from Seville who accompanied Christoffa Corombo (a.k.a. Christopher Columbus) during his second voyage to the New World in 1493. Columbus had come ashore on the island he would name Santa María de Guadalupe. As the landing party entered a small village, the local inhabitants fled in terror, leaving behind everything they owned. It was a response that had taken them a little over a year to develop.

During his first voyage in 1492, Columbus referred to all of the native people as *indios*, but by a year later a distinction had been made between the peaceful Arawaks (also called Taínos) and another group known to the locals as the Caribes (or more commonly, Caribs). What Columbus would never know was that the indigenous inhabitants were actually a diverse assemblage that had been living on the islands for hundreds of years. Their ancestors had set out from coastal Venezuela, where the out-flowing currents of the Orinoco River carried the migrants into the open sea and far beyond. At each island stop, these settlers developed their own cultures and customs, so that by the time the Spaniards arrived, the entire Caribbean island chain had already been colonized, with settlements extending as far north as the Bahamas.

Columbus, though, cared little about local customs or history. Instead he noted that the Arawaks were gentle and friendly, and he wasted little time in passing this information on to his royal backers in Spain. "[The Arawaks] are fitted to be ruled and to be set to work, to cultivate the land and do all else that may be necessary."

This somewhat-less-than-friendly response led the locals to initiate some self-preservation-related finger pointing, designed perhaps to send their new pals off on a quest to enlighten somebody else. Although no one is quite sure who was doing the translating, soon after his initial arrival, the Arawaks reportedly told Columbus that the Caribs inhabited certain of the southern islands, including those that would eventually be called St. Vincent, Dominica, Guadeloupe, and Trinidad. Columbus was informed that the Caribs were not only infamous for brutal raids against their peaceful neighbors, but also for the annoying habit of eating their captives. But pillaging and people-eating weren't the Caribs' only vices. Every year they took a break from their mayhem-related jobs in order to meet up with a tribe of warrior women. These fighting females were reportedly "fierce to the last degree, strong as tigers, courageous in fight, brutal and merciless."

With more than a fleeting resemblance to a race of fictional characters dreamed up by the Ancient Greeks, these warrior women lived on their own island (Martinique) and killed any men they encountered . . . except, that is, for the Caribs, who got a yearly invite to drop by for some feasting and debauchery. Possibly the invitations stemmed from the fact that the Caribs were renowned for their cooking ability—preparing their viands by smoking them slowly on a wooden platform. It was a setup the Spanish began referring to as a *barbacoa*. After manning the grill and servicing the gals, the Caribs returned home, taking with them any newborn males who had shown up nine months after the previous year's party. Female babies would, of course, remain behind with their moms to be raised as warriors.

In retrospect, it is difficult to determine where Arawak tall tales ended and Columbus's vivid and self-serving imagination kicked in. What is known is that European history and folklore were already rich with references to encounters with bizarre monsters and strange human races. Although most of the stories emerging from the New World were greeted with enthusiasm back in Seville, some of Columbus's patrons expressed skepticism after hearing that the Caribs also hunted with schools of fish. These had been trained to accept tethers and dispatched with instructions to latch on to sea turtles, which could then be reeled in for butchering.[14]

Easier to accept, perhaps, were Columbus's claims that some Caribs had doglike faces, reminiscent of the *Cynocephali* described nearly 1,400 earlier by Pliny the Elder, the Roman author and naturalist. Still other New World locals were said to possess a single, centrally located eye or long tails—appendages that necessitated the digging of holes by their owners so that they might sit down. These creatures were considered anything but a joke, since as late as 1758, Linnaeus's opus *Systema Naturae* listed three species of man: *Homo sapiens* (wise man), *Homo troglodyte* (cave man), and *Homo caudatus* (tailed man).

But whether or not these strange savages had tails (and even if they were supported by trained fish and Amazonian girlfriends), plans were soon being formulated to pacify the Caribs, who were

---

14 In Northern Australia, East Africa, and the Indian Ocean, some cultures do employ a family of sucker-backed fish called remoras (Echeneidae) to hunt for sea turtles. Remoras are renowned for attaching themselves to larger fish as well as turtles. The original behavior is a form of commensalism—a relationship in which one species (the remora) obtains a benefit (in this case protection and food dropped by the host) while the other species gains nothing but isn't harmed.

now being referred to as Canibs. According to scholars, the transition from Carib to Canib apparently resulted from a mispronunciation, although in light of stories describing locals as having canine faces, I agree with Yale professor Claude Rawson that "Canib" may also be a degenerate form of *canis*, Latin for "dog." Eventually *canib* became the root of "cannibal," which replaced *anthropophagi*, the ancient Greek mouthful previously used to describe people-eaters.

But whatever the locals were called, and however the term *cannibal* may have originated, the first part of Columbus's grand plan centered on relieving them of the abundant gold he "knew" they had in their possession. One reason for Columbus's certainty on this point was the commonly held belief that silver formed in cold climates while gold was created in warm or hot regions. And considering the heat and humidity of the New World tropics, this could only mean that there would be plenty of the shiny stuff around.

Unlike his first voyage, which consisted of three ships and 120 men, Columbus's second visit to the New World had the look of a military occupation force. Accompanying him were 17 ships and nearly 1,500 men, many of them heavily armed. Although he had begun to look at slave raiding as a means to finance his voyages, his prime directive was to find gold—lots of it. To facilitate the collection of what Columbus assumed would be a massive treasure, he levied tribute on those living in regions like El Cibao in what is now the northern part of the Dominican Republic. His orders stated that every male between 14 and 70 years of age was to collect and turn over a substantial measure of gold to his representatives every three months. Those who failed at what quickly became an impossible task had their hands hacked off. Anyone who chose to flee was hunted down—the Spaniards encouraging their vicious war-dogs to tear apart any escapees they could track down.

In the end, very little of the precious metal was turned in. Presumably the island residents, under the very real threat of losing their limbs or being eaten alive by giant dogs, quickly ran through any gold they might have had on hand. Since it played only a small role (or no role at all) in their traditions, in all likelihood the locals just didn't know where to find it—especially in the quantities demanded by the Spanish invaders.

Deeply disappointed at the meager results, Columbus penned

a letter to his royal supporters in Spain in May 1499. In it he wondered "why God Our Lord has concealed the gold from us." There is no record of a response but Columbus soon refocused his efforts toward the collection of a resource that *was* available in great supply—humans.

In 1503, this bloodthirsty new take on the exploration of the New World got a significant boost when the self-proclaimed Admiral of the Ocean Sea received a royal proclamation from Queen Isabella. In it she stated that those locals who did not practice cannibalism should be free from slavery and mistreatment. More significantly, though, she also instructed Columbus and his men about what they *could* do to them if they were determined to be cannibals:

> If such cannibals continue to resist and do not wish to admit and receive to their lands the Captains and men who may be on such voyages by my orders nor to hear them in order to be taught our Sacred Catholic Faith and to be in my service and obedience, they may be captured and are taken to these my Kingdoms and Domains and to other parts and places and be sold.

This new position was given even more support by the Catholic Church several years later, when Pope Innocent IV decreed in 1510 that not only was cannibalism a sin, but that Christians were perfectly justified in doling out punishment for cannibalism through force of arms.

What happened next was as predictable as it was terrible. On islands where no cannibalism had been reported previously, maneating was suddenly determined to be a popular practice. Regions inhabited by peaceful Arawaks were, upon reexamination, found to be crawling with man-eating Caribs, and very soon the line between the two groups was obliterated. "Resistance" and "cannibalism" became synonymous, and anyone acting aggressively toward the Europeans was immediately labeled as a cannibal.

In an effort to organize the cannibal pacification efforts, Rodrigo de Figueroa, the former governor of Santo Domingo (now the Dominican Republic), was given the job of making judgments on the official classification of all the indigenous groups encountered by the Spanish during their takeover. Testimonials and other "evidence" were used to place the cannibalism tag on island populations—and by a strange coincidence, the designations seemed to change with the priorities of the Spanish for the islands in question. Trinidad, for example, was declared a cannibal island in 1511, but the ruling was changed in 1518. Rather than relating to concerns over the welfare of the local people, though, the reclassification came about because of reports of gold in Trinidad and the Spaniards' desire to maintain the local population for use in mining operations. It was more than coincidental, then, that once the Spanish mining efforts on Trinidad failed to produce any gold, reports began filtering in that the locals were cannibals after all. Soon after, the order was given to colonize Trinidad and to depopulate it of its remaining man-eating inhabitants. As a result, the pre-Columbian indigenous population in Trinidad (estimated to be somewhere between 30,000 and 40,000 individuals) dropped to half that number within 100 years.

Even in places that hadn't initially been designated as cannibal islands, populations dropped precipitously as the locals were either hauled off to toil as slaves, were murdered, or died from newly arrived diseases like measles, smallpox, and influenza (the latter may have been a form of swine flu carried by some pigs that Columbus had picked up on the Canary Islands during the early part of his second voyage). According to historian David Stannard, "Wherever the marauding, diseased, and heavily armed Spanish forces went out on patrol, accompanied by ferocious armored dogs that had been trained to kill and disembowel, they preyed on the local communities, already plague-enfeebled, forcing them to supply food and women and slaves, and whatever else the soldiers might desire."

The diseases the Spaniards carried (the precise identities of which are still debated) spread with alarming speed through local communities, killing inhabitants in numbers that, according to one writer at the time, "could not be counted." Stannard believes that by the end of the 16th century, the Spanish had been directly or indirectly responsible for the deaths of between 60 and 80 million indigenous people in the Caribbean, Mexico, and Central America. Even if one were to discount the millions of deaths resulting from diseases, this would still make the Spanish conquest of the New World the greatest act of genocide in recorded history. These types of numbers, which are subject to considerable academic debate, are often overlooked during Columbus Day parades and related festivities.[15]

---

15 Political scientist Rudolf Rummel estimates that, excluding military battles and unintentional (e.g., disease-related) deaths, European colonization killed between 2 and 15 million indigenous Americans, with the vast majority of deaths taking place in Latin America.

In the end, tall tales, especially those with bestial or cannibalistic angles, effectively dehumanized the islanders. Not only did this serve to justify Spain's rapidly evolving slave-raiding agenda, but it also established a mindset toward the locals that came to resemble pest control. Leaving behind neither pyramids nor stone glyphs, the indigenous cultures of the Caribbean have all but disappeared.

# 10: Bones of Contention

*I do not think it is an exaggeration to say history is largely a history of inflation, usually inflations engineered by governments for the gain of governments.*

—Friedrich August von Hayek, Austrian economist (1899–1992)

Peter Martyr (1457–1526) was an Italian cleric who never set foot in the New World. Nevertheless, *De Orbe Novo* (*On the New World*), which was published in 1530, was an epic depiction of the first eight decades of Spanish rule in the West Indies. It became one of the most influential and popular works ever written on the subject. Without the benefit of firsthand knowledge, Martyr obtained the information for his book from interviews conducted with sailors, clergymen, and others returning from overseas. In Book One, which detailed the voyages of Christopher Columbus, the author included a section on the notorious cannibal hut described by Dr. Chanca during Columbus's second voyage (and whose quote opened the previous chapter). Martyr, however, appears to have taken a bit of creative license with the physician's account, expanding the incident and giving it a truly horrific tone. Instead of the single hut described by Chanca, there were multiple dwellings, each outfitted with a kitchen in which

> birds were boiling in their pots, also geese mixed with bits of human flesh, while other parts of human bodies were fixed on spits, ready for roasting. Upon searching another house the Spaniards found arm and leg bones, which the cannibals carefully preserve for pointing their arrows; for they have no iron. All other bones, after the flesh is eaten, they throw aside. The Spaniards discovered the recently decapitated head of a young man still wet with blood.

Clearly, Martyr was instrumental in dehumanizing the Caribs, describing them as savages who treated their fellow islanders in the much the same way Europeans might treat sheep or cattle. Additionally, in keeping with his pro-Columbus stance, Martyr also used the threat of cannibals (now described as having nearly supernatural powers) as a thinly veiled justification for the overt military theme of Columbus's third voyage, an expedition that became, in effect, a New World troop surge:

> The inhabitants of these islands (which, from now on we may consider ours), women and men have no other means of escaping capture by the cannibals, than by flight. Although they use wooden arrows with sharpened points, they are aware that these arms are of little use against the fury and violence of their enemies, and they all admit that ten cannibals could easily overcome a hundred of their own men in a pitched battle.

So was there any *real* cannibalism going on in the Caribbean when Columbus arrived? Oxford-trained anthropologist Neil Whitehead suggests that while many reports of the behavior are examples of "imperial propaganda," there are several reasons to think that the Caribs and other Amerindian groups did practice forms of ritualized cannibalism. Whitehead's rationale is that, in addition to the self-serving allegations of man-eating from Columbus and his men, reports from other Spaniards placed Amerindian cannibalism into social contexts—as funerary rites or rituals related to the treatment of enemies slain during battle. For example, in the 17th century, Jacinto de Caravajal wrote, "The ordinary food of the Caribs is cassava, fish or game . . . they eat human flesh when they are at war and do so as a sign of victory, not as food."

According to anthropologists, ritualized cannibalism can be differentiated into two forms: exocannibalism and endocannibalism. *Exocannibalism* (from the Greek *exo*—"from the outside") refers to the consumption of individuals from outside one's own community or social group, while *endocannibalism* (from the Greek *endo*—"from the inside") is defined as the ritual consumption of deceased members of one's own family, community, or social group.

With regard to exocannibalism, a number of historical accounts claim that the Caribs consumed their enemies—those killed in battle, taken prisoner, or captured during raids. The belief was that this form of ritual cannibalism was a way to transfer desired traits, like strength or courage, from the deceased enemy to themselves.

In other times and places, exocannibalism has been used as a way to both terrorize an enemy and feed the hungry. In the 1960s, anthropologist Pierre Clastres lived with the Ache of Paraguay and claimed that one of the four groups that he studied ate their enemies. Similar claims have been made about the Tupinambá of eastern Brazil, most famously by Hans Stadin, a 16th century German shipwrecked while serving as a seaman on a Portuguese ship. In his 1557 book, *True Story and Description of a Country of Wild, Naked,*

*Grim, Man-eating People in the New World, America*, Stadin, who reportedly spent a year in captivity before escaping, described raids in which the Tupinambá killed and ate everyone they captured (except, apparently, him).

In the Pacific Theater during World War II, Allied prisoners of war described numerous instances in which their Japanese captors tortured and then ate their prisoners. Presumably with their supply routes interrupted by Allied submarines and bombing raids, the Japanese were on such short rations that they resorted to cannibalism. In postwar tribunals, survivors testified that their captors acted systematically, selecting one individual each day and hacking off limbs and flesh while they were alive and conscious. American soldiers also became even more insistent about removing the bodies of their fallen comrades from the battlefield after it was discovered that the Japanese sometimes sliced off pieces of the dead with bayonets—a gory ritual some Americans began to practice as well.

The most famous wartime incident of starvation-related exocannibalism was the Chichi Jima Incident, in which Lt. Gen. Yoshio Tachibana ordered his starving men on the island of Chichi Jima to execute a group of downed American airmen who had been captured after carrying out a bombing raid. Medical orderlies were then instructed to cut the livers from the bodies, and the organs were cooked and served to the senior staff. Tachibana and several others were arrested after the war, but since cannibalism was not listed as a war crime, they were actually convicted and hanged for preventing the honorable burial of the prisoners the officer and his men had eaten. Later was it revealed that an American submarine had recovered one of the nine downed fliers, thus saving him from a similar fate at the hands of the starving Japanese. The lucky man's name was Lt. George H. W. Bush.

There is no such element of terror involved in the practice of endocannibalism, although it can overlap with some aspects of exocannibalism in that body parts (in this case, from relatives or group members) are consumed for reasons that include transferring the spirits of the dead or their traits into the bodies of the living. Anthropologists have proposed that, much like Christian burial rituals or the administration of Last Rites, endocannibalism was undertaken by some groups in order to facilitate the separation of the deceased's soul from its body. The Melanesians (those societal groups living in Fiji, the Solomon Islands, Vanuatu, and Papua New Guinea) reportedly practiced a form of mortuary cannibalism for this reason, consuming small tidbits from the bodies of their deceased relatives. This form of ritual cannibalism will be examined in detail in an upcoming chapter.

Anthropologist Beth Conklin studied the Wari' from the western Amazonian rainforest of Brazil. She reported that until the 1960s, the Wari' consumed portions of human flesh as well as bone meal mixed with honey. Having conducted extensive interviews with Wari' elders, she said that the "Wari' are keenly aware that prolonged grieving makes it hard for mourners to get on with their lives." With the corpse being the single most powerful reminder of the deceased, the Wari' believed that consuming the body eradicates it once and for all. Beliefs or not, though, they were forced by missionaries and government officials to abandon their funerary rites and to bury their dead in what these strangers believed to be the civilized manner. Conklin said that this was a ritual the Wari' found to be particularly repellent, since they considered the ground "cold, wet and polluting" and that "to leave a loved one's body to rot in the dirt was disrespectful and degrading to the dead and heart-wrenching for those who mourned them."

Getting back to the question of whether or not the Caribs were cannibals, several years ago, during a trip to Trinidad, I met Cristo Adonis, the spiritual leader of the Trinidadian Amerindians. One of the first things I learned was that there were very few of his people left (with no full-blooded individuals among the roughly 600 surviving members). I quickly noted that Adonis avoids using the European-assigned names Carib and Arawak, which he considers slang terms. Instead he refers to his ancestors as the Karina and Locono people. Mr. Adonis told me that his people did practice both endocannibalism, related to religious practices, and exocannibalism, as a way to gain power from their defeated enemies. His evidence for this claim derives solely from ethnohistorical accounts passed down over hundreds of years. I told him that might make some folks skeptical.

"Why," he asked me, "would indigenous historians pass on stories about their ancestors practicing cannibalism if the stories weren't based on actual customs?" Actually, I can think of some potential reasons for claiming one's ancestors were cannibals (e.g., to instill fear in their enemies), but anthropologist Neil Whitehead also thinks that the Caribs were man-eaters, although I found his rationale to be open to debate. To back this up his claim, Whitehead offers accounts by non-Spanish writers describing the Amerindians they encountered as having carried out ritual cannibalism. Whitehead argues that since the English, French, and Dutch were enemies of the Spanish, they would have wanted to develop alliances with the Amerindians. Since the non-Spaniards were presumably on friendlier terms with the locals, they would have been in a better position to observe and report on the actual cannibalistic behavior of their native allies.

Arguing against Carib cannibalism, perhaps, is the fact that the documentation by non-Spaniards regarding the behavior contains some seriously fanciful descriptions. For example, alongside his descriptions of anthropophagy, Sir Walter Raleigh wrote about some indigenous peoples having their heads located within their chests and their feet pointing backwards, the latter a characteristic that made them "very difficult to track." (See illustration on page 103.)

Interesting for another reason entirely is the most famous piece of Columbus-related circumstantial evidence: Dr. Chanca's account of the recovery of "four or five bones of human arms and legs" in a hastily abandoned hut. In reality, the good doctor never saw the scene he wrote about in 1493 because he was not part of the landing party. This might come as a surprise, though, if you read Chanca's work, since his repeated use of the word "we" gives the impression that he had experienced the horrors of the "cannibal hut" firsthand.

Though not a firsthand witness, Dr. Chanca was a strong supporter of Columbus, and because of his professional status, his written accounts carried tremendous weight. Historically, his letters make up much of what we know (or thought we knew) about the Admiral's second voyage to the New World. Another factor to consider is that Chanca's description of the cannibal hut was sent back to Spain accompanied by a letter from Columbus, requesting that the doctor's salary be increased substantially. Since Columbus was already using the cannibal angle to justify his attempts to pacify and enslave the local residents, what are the odds that Dr. Chanca would have penned an accompanying document contradicting the Admiral's description of the Caribs as subhuman eaters of men? This conflict of interest raises serious doubts as to whether Chanca's account can be considered unbiased reporting.

But even if the events described by Chanca did take place, the bones Columbus and his men collected from the infamous hut were more likely part of a funerary ritual rather than proof of cannibalism. According to historians and anthropologists, rather than burying their departed ancestors, some Amerindians preserved and worshipped their bones. In 1828, author and historian Washington Irving pointed out that during Columbus's first voyage, when human bones were discovered in a dwelling on Hispaniola, they were taken to be relics of the dead, reverently preserved. On Columbus's second visit, however, when bones were found in a hut presumably inhabited by Caribs, the finding became incontrovertible evidence of cannibalism.

At best, then, Dr. Chanca's letter provides a brief, secondhand account of what may or may not have been the aftermath of cannibalism by the inhabitants of a single hut on the island of Guadeloupe. Meager evidence? Certainly, but the story gained far greater significance as additional authors wrote about the incident. In what would become a blueprint for cannibal tales throughout history, descriptions of the practice were penned decades or even centuries after the actual event and without the input of additional witnesses.

It mattered little whether cannibalism took place in the New World or not, though, since most authors who wrote on the topic had already decided that it had. Historians and writers alike picked up on secondhand, unsubstantiated, and often fabricated stories, added a splash of red, then reported the tales as facts. In doing so they may have enthralled their audiences with the bravery of European explorers, but they did a terrible disservice to history and to the indigenous people who became less and less human with each exaggerated account. As a result, readers—both casual and scholarly—were subjected to a 500-year-long indoctrination period during which they heard little if anything about the genocidal mistreatment of native populations, or even the sociological significance of cannibalism (if the practice did occur). Far more likely, they would come away believing that Columbus and the other European explorers had fought off hordes of cannibalistic subhumans, thus sparing many a grateful savage the horrors of the cooking pot. From the New World to Africa, Australia, and the Pacific islands, cannibalism was generally perceived to be a widespread phenomenon and it would be the role of good Christians—explorers and the missionaries who invariably followed them—to take control of the situation and thus put an end to this most horrific of human behaviors.

For the most part, this public mindset concerning ritual cannibalism remained until 1979. It was then that Stony Brook University anthropology professor William Arens initiated what became a loud and serious debate over the validity of cannibalism as a social practice. In his book *The Man-Eating Myth*, Arens argued that, aside from some well-known instances of starvation-induced cannibalism, there was absolutely no proof that ritualized cannibalism had ever existed in any human culture. Instead, according to the anthropologist, the idea of cannibalism had become a handy symbol for unacceptable behavior practiced by "Others"—a broad and malleable category of evildoers that included enemies, followers of non-Christian religions, and any groups determined to retain their "uncivilized" customs.

"The most certain thing to be said is that all cultures, subcultures, religions, sects, secret societies and every possible human association have been labeled as anthropophagic by someone." Essentially, then, Arens asserted that colonial groups had been guilty of making false accusations of cannibalism against native populations across the globe and throughout history. With Christopher Columbus acting as a poster boy, applying the cannibal tag justified the condemnation and, if necessary, the eradication of anyone accused of engaging in this ultimate of taboos—a practice whose validity (Arens was quick to point out) was always unsupported by anything resembling firsthand evidence.

The reaction to Arens's incendiary book was swift and mostly negative. His colleagues referred to his man-eating-myth hypothesis as "unsophisticated" and "dangerous," and that it "does not advance our knowledge of cannibalism." Some critics also took the opportunity to attack Arens personally, with the most extreme assault coming from anthropologist Marshall Sahlins, who claimed that Arens had proposed his "outrageous theory" for the sole purpose of generating controversy in an effort to sell books. Sahlins took his accusations over the edge by comparing Arens to a Holocaust denier, a stance that even those who believed ritual cannibalism to be a commonplace occurrence found difficult to fathom. One anthropologist, who believed that cannibalism did occur, pointed out that Holocaust victims had been murdered because, like the Caribs, they had been labeled as Others and, as such, they became perfect candidates for extermination.

I've found myself agreeing with much but certainly not all of Arens's hypothesis, in part because of the brutal pounding colonial invaders doled out to indigenous groups for centuries. On the other hand, my investigation into ritual cannibalism has led me to conclude that there is plenty of evidence to support the stance that some cultural groups practiced cannibalism and that they did so for a variety of reasons. As for the claims of Carib cannibalism, though, the fact remains that beyond the second- and third-hand accounts, there isn't a shred of physical evidence, nor is there any indication that Columbus or his men ever witnessed man-eating firsthand.

I MET DR. Arens on a bright, sunny day in his office at Stony Brook University. He seemed apprehensive at first, but after telling him I had little interest in the sensational aspects of cannibalism and that I had an open mind regarding his own assertions on the topic, he began to open up.

He explained that when anthropologists address the more recent claims of ritual cannibalism, much of the evidence they present comes from interviews they or their associates have had with tribal elders—none of whom practice cannibalism anymore. As Arens is only too happy to point out, these former cannibals all seem to have ceased their man-eating behavior just before the arrival of the anthropologists. I gathered that this was too much of a coincidence for him to accept.

In response to this very point, Beth Conklin (who worked with the Wari' in Brazil) argued previously that the reason anthropologists never observe cannibalism in person is because they are "seldom the first outsiders to set foot in a newly contacted society." This role, she claimed, is filled by "missionaries, government agents, frontiersmen or traders, who see cannibalism as something to terminate immediately." Thus, according to Conklin and others, cannibalism either ceases altogether or goes underground *before* it can be studied firsthand.

Conklin also raised her own concerns about the cannibal denial debate:

> Like the many priests, missionaries, colonial officers, and others who considered cannibalism antithetical to what it means to be human, scholars who insist that all accounts must be false seem to assume that cannibalism is by definition a terrible act. They appear blind to the possibility that people different from themselves might have other ways of being human, other understandings of the body, or other ways of coping with death that might make cannibalism seem like a good thing to do.

In other words, just because we consider cannibalism an ultimate taboo, why should the members of other cultural groups necessarily feel the same way?

I asked Arens about Conklin's statement.

"I think it's nonsense," he replied.

"But couldn't there be a group of isolated people who grew up without the influences that lead Westerners to believe that cannibalism is a bad thing?"

Arens threw me a dismissive wave. "I don't think that any group of people grow up isolated or innocent of what's going on around them, and I don't believe that one group does something that's not pretty pervasive among the species."

He continued. "But if you see people eating each other, then you have to accept that they do it. And although I'd be disappointed to have to accept that, I would accept it!"

*Yeah right*, I thought, retaliating with a respectful but dismissive wave of my own. I found it hard to fathom that a man who had pretty much made a career of rejecting the concept of ritual cannibalism might be converted so quickly.

"Honestly, I'd accept it," Arens assured me. "My problem is that no one ever sees it. Therefore, the pattern [of observed behavior] is not eating people, but *assuming* people eat people and never actually seeing it. The example of the Wari', that's a problem because no one has seen the Wari' practicing cannibalism. So how about the Bongo Bongo? No one's ever seen the Bongo Bongo do it, either. So you end up going around the world discussing something that no one's ever seen."

Anthropologist Jerome Whitfield has been working on ritual cannibalism and its pathological consequences for several decades, principally in Papua New Guinea. I'd been corresponding with Whitfield and related to him my conversation with Arens. He responded via email. "Endocannibalism has been practiced and witnessed by thousands of indigenous people throughout the world. How come a white ethnocentric anthropologist, who has never spoken to these people, or been in their country, can say what they do or why they do something?"

Whitfield went on. "Interestingly, there have been public apologies in Fiji, Vanuatu, and Papua New Guinea by citizens of these countries for the killing and exocannibalism of missionaries. Are they making it up like the Wari'?" To Whitfield, it sounded like Arens believes there is "a huge conspiracy with massive resources that wants to mislead the world into thinking that endocannibalism was an established practice." And as for a reliance on witnesses other than themselves, Whitfield wrote that, "All anthropologists rely on informants to explain what is happening, even if they witness the event themselves."

So did ritual cannibalism ever take place? Most anthropologists who've investigated the topic seem to think so. But is there any evidence beyond the ethnohistorical accounts and, if so, does it still take place today? As I would learn, to millions of people, the answer is apparently yes.

# 11: Cannibalism and the Bible

*I had to eat a piece of Jesus once in a movie.*

—John Lurie (personal communication),
costar of Martin Scorsese's *The Last Temptation of Christ*

There is another form of ritual cannibalism whose origins are fascinating and whose details may strike many readers as being uncomfortably close to home.

Descriptions of cannibalism in the Bible fall into two distinct categories. In the Old Testament, the behavior was undertaken by the starving inhabitants of the besieged cities of Jerusalem and Samaria. There's no physical evidence that these events actually occurred (although that doesn't mean that they didn't), but since we'll be covering the topic of survival cannibalism in an upcoming chapter, we won't be stopping here.

The second type of cannibalism is found in the New Testament and relates to the literal or symbolic consumption of Jesus Christ's body and blood during the celebration of the Eucharist—the Christian commemoration of the Last Supper. Considering the paramount importance this ceremony has for all Christians, and in light of differing belief systems that exist throughout Christianity, it's no surprise that there are disagreements concerning the interpretation of the Eucharist. One aspect shared by the vast majority of Christians, however, is a lack of awareness that this particular form of ritual cannibalism led to the torture and death of thousands of innocent people.

The following are two of the most famous passages from the New Testament.

> Now as they were eating, Jesus took bread, and when he had said the blessing he broke it and gave it to the disciples. "Take it and eat," he said, "this is my body." Then he took a cup, and when he had given thanks he handed it to them saying, "Drink from this, all of you, for this is my blood, the blood of the covenant, poured out for many for the forgiveness of sins."
> —Matthew 26:26–28

> Jesus replied to them: In all truth I tell you, if you do not eat the flesh of the Son of man and drink his blood, you have no life in you. Anyone who does eat my flesh and drink my blood has eternal life, and I shall raise that person up on the last day. For my flesh is real food and my blood is real drink. Whoever eats my flesh and drinks my blood lives in me and I live in that person.
> —John 6:53–56

One way to interpret these passages is that Jesus was using a metaphor to convey a concept to his followers. It was certainly something he had done before, since surely even the dimmest of Jesus's supporters hadn't taken him literally when he said, "I am the gate" (John 10:9) or "I am the true vine" (John 15:1). In fact,

Jesus's lesson to his disciples during the Last Supper is one of those seemingly rare instances where even evangelical Christians appear to bend their own rules regarding literal translation. For example, the same fundamentalists who believe that Jonah was swallowed by a fish and survived for three days within its belly also believe that the wine and host they consume during Holy Communion are only symbols of the body and blood of Jesus Christ. Strangely, though, the leaders of several major Christian religions (including Catholicism) do not support this host-and-wine-as-symbols interpretation, at least not technically. Here's how that disagreement came about.

In light of developments resulting from the first four Crusades (e.g., the capture of Constantinople and large parts of the Byzantine Empire), Pope Innocent III summoned over 400 bishops and many lesser ecumenical leaders to attend the Fourth Lateran Council in 1215. Representative rulers from Europe and the Levant were also invited (the latter in reference to an area now made up of Lebanon, Israel, Jordan, the Palestinian territories, Syria, and Iraq). During the meeting, there was apparently little discussion between the Pope and the council attendees. Instead, the pontiff presented a list of 71 papal decrees, which served notice to all present that the Pope's powers, as well as those of the Roman Catholic Church, had just been expanded. Among proclamations forbidding the founding of new religious orders, strengthening papal primacy, and regulating and restricting Jewish communities, was a decree that spelled out the concept of transubstantiation.

As pronounced by Innocent III, from that moment on, the faithful would be required to believe that the consecrated elements in the Eucharist (i.e., the bread and wine) were literally changed into the actual body and blood of Jesus Christ. "His body and blood are truly contained in the sacrament of the altar under the forms of bread and wine, the bread and wine having been changed in substance, by God's power, into his body and blood."

If the council attendees had any gripes about these new decrees they apparently kept them to themselves. During the 16th century, however, the interpretation of biblical passages like those describing the Last Supper became pivot points for the controversies that arose between the Catholics and Protestants. In that regard, Martin Luther (leader of the Protestant Reformation) seemed to have more than a little problem with the whole idea of transubstantiation, beginning with the fact that the term did not appear in any biblical scriptures. Apparently, Hildebert of Lavardin, Archbishop of Tours, had coined the term, from the Latin *transsubstantiatio*, around 1079 CE. In 1520, though, Martin Luther referred to it as "an absurd and unheard-of juggling with words" and stated, "the Church had the true faith for more than twelve hundred years, during which time the holy Fathers never once mentioned this transubstantiation—certainly, a monstrous word for a monstrous idea."

A decade later, the Incan king Atahualpa took issue with the concept of transubstantiation. In their entertaining book *Eat Thy Neighbor*, authors Daniel Diehl and Mark Donnelly recount the story of what took place after the capture of Atahualpa by conquistador Francisco Pizarro in 1533. Threatened with execution unless he converted to Christianity,

> Atahualpa said he bowed to no man and told the Spanish exactly what he thought of their religion. His people, he said, only sacrificed their enemies to their gods and certainly did not eat people.

> The Spanish, on the other hand, killed their own God, drank his blood and baked his body into little biscuits which they sacrificed to themselves. He found the entire practice unspeakable. The Spanish were outraged and had Atahualpa publicly executed.

Unfortunately, other accounts of this incident offer a somewhat less cinematically heroic end to Atahualpa's story. In an alternate version, the captured Incan king converted to Catholicism and was given the name Juan Santos Atahualpa. His fellow Catholics then celebrated Juan's baptism by having him strangled with a garrote.

Back on the other side of the wafer, Roman Catholic leaders not only adopted the concept of transubstantiation but during the Eastern Orthodox Synod of Jerusalem (a famous get-together in 1672), they took a moment to thumb their collective noses at the upstart Protestants:

> In the celebration of [the Eucharist] we believe the Lord Jesus Christ to be present. He is not present typically, nor figuratively, nor by superabundant grace, as in the other Mysteries, nor by a bare presence . . . as the followers of Luther most ignorantly and wretchedly suppose. But truly and really, so that after the consecration of the bread and of the wine, the bread is transmuted, transubstantiated, converted and transformed into the true body itself of the Lord . . . and the wine is converted and transubstantiated into the true blood itself of the Lord."

Even as recently as 1965, Pope Paul VI made it clear that as far as he and the Roman Catholic Church were concerned, with regard to transubstantiation, their stance had not changed in the 400 years since the Council of Trent, one of the Church's most important ecumenical councils.

But how many of the modern faithful ever think about the concept of transubstantiation when they're taking communion? And similarly, how many Catholics are worried about being labeled heretics if they don't really believe that they're performing an act of theophagy as they consume the wine and wafer? Apparently not many. In fact they seem to take the same kind of "nod-nod, wink-wink" approach to transubstantiation as they do toward not eating meat on Fridays or, in the case of my relatives under the age of 85, the church's ban on any form of birth control beyond the rhythm method—which several of them refer to as Vatican Roulette.

This laid-back attitude, however, was definitely not present in the years that followed the Fourth Lateran Council of 1215. Backed by a pope who decreed that their communion bread was the actual body of Jesus, church officials began persecuting those whom they suspected of abusing it. Beginning some 30 years after Pope Innocent's decree concerning transubstantiation, faithful Catholics started rounding up and executing Jews for the crime of "torturing the host."

How, you might ask, did the accusers know that their hosts were being desecrated? Apparently, unimpeachable witnesses came forward, claiming to have seen the communion bread bleeding. I'll get back to that one, but first up is the question of why the Jews were being blamed for these awful crimes. The answer appears to be that while there wasn't a shred of evidence that they were involved in unspeakable, host-related acts, at the time everyone did seem to agree that the Jews hated Jesus—in fact they had been responsible for his death, hadn't they? Maybe, the accusers reasoned, the Jews

Adapted from a 15th-century German woodcut depicting host desecration by the Jews of the Bavarian town of Passau in 1477. The hosts are stolen and brought to a temple where they are pierced with a dagger during some unspecified Jewish ritual carried out in the presence of a Torah. Eventually, the hosts are rescued in a commando-like raid and the communion wafers are shown to be holy. The guilty Jews are arrested. Some are beheaded, others tortured with hot pincers. Next, the entire Jewish community has their feet put to the fire before being driven out of town (or to their death). In the end, the good Christians kneel and pray.

were reenacting the Messiah's crucifixion or using the host as part of their own nefarious rituals. Rumors had begun circulating that Jews were applying the blood that flowed from the host to their faces, to give their cheeks a rosy appearance. Other host-conspiracy buffs suggested that the villains were using the Savior's blood to rid themselves of the *foetor Judaicus* ("Jewish stink").

And so it came to pass that, in the complete absence of anything resembling evidence, Jews were rounded up, coerced, and tortured—after which many of them fessed up to the crimes they hadn't really committed. But whether they confessed or not, those found guilty of defiling the sacrament were subjected to additional torture before being burnt at the stake, beheaded, or dispatched in some equally gruesome manner. Additionally, their families, as well as any neighbors brazen enough to have committed the crime of "living nearby" often accompanied them to their deaths. These ghastly practices continued for nearly 400 years in Jewish communities all across Europe, with massacres taking place in Germany, France, Austria, Poland, Spain, and Romania.

At some point the execution of Jews for crimes against baked goods ended. Unfortunately, the reason for the cessation of these pogroms had nothing to do with Christians coming to their senses about just how badly they had been acting. Instead, it had everything to do with finding a new group—witches—to persecute for similar crimes. So before you could say, "Got a match?" witches were being burned alive for having a weird mole, or trying to procure red hosts for their Black Mass, or associating with communists. (All right, that last one didn't happen until the 1950s.)

But what about those bleeding hosts? Were medieval witnesses just making that stuff up as an excuse to get rid of a group they despised? Or maybe these folks had simply imagined the ruby-stained bread? There is, however, an intriguing alternative hypothesis. In 1994, Dr. Johanna Cullen, at George Mason University in Virginia, came up with an explanation for bleeding hosts that was neither mystical nor mental. It was instead, microbiological. *Serratia marcescens* is a rod-shaped bacterium and common human pathogen frequently linked to both urinary tract and catheter-associated infections. The ubiquitous microbe can also be found growing on food like stale bread that has been stored in warm, damp environments. For this story, the key characteristic of *S. marcescens* is that it produces and exudes a reddish-orange pigment called prodigiosin, a substance that can cause the bacterial colonies to resemble drops of blood. Clinically, prodigiosin has been shown to be an immunosuppressant with antimicrobial and anti-cancer properties and it's likely that these germ-killing properties protect *Serratia* colonies from attack by bacteria, protozoa, and fungi, in much the same way that the *Penicillium* mold produces an antibacterial agent that has been co-opted for use by humans. In the 15th century, though, *Serratia* colonies growing on the host may very well have been mistaken for blood.

The work of another researcher, Dr. Luigi Garlaschelli, backed up Dr. Cullen's findings. The renowned organic chemist and part-time debunker of reputed miracles like weeping or bleeding statues examined various food items that were said to have bled spontaneously. To determine whether the "blood" was real or not, Garlaschelli tested the items for the presence of hemoglobin, the oxygen-carrying pigment that gives vertebrate blood its red color. In the end, the tests revealed no hemoglobin but plenty of contamination by *S. marcescens*, and the Italian chemist further demonstrated the likely origin of the bleeding hosts by culturing the bacterium on slices of ordinary white bread.

Quite possibly, then, a common microbe contaminated the bleeding hosts of the Middle Ages, which is actually kind of amusing until you realize how many thousands of innocent people were murdered because of this tragic bit of ignorance and misinterpretation.

A final word on the relationship between transubstantiation and cannibalism concerns the Uruguayan survivors of the Old Christians Rugby Club, who employed what became known as the "communion defense" to justify the incidents of cannibalism that took place after their 1972 plane crash in the Andes. Soon after the 16 survivors returned to civilization, positive public opinion over their plight took a nosedive after it was revealed that the men had remained alive for 72 days by consuming the bodies of the dead. Not long after their rescue, and with their hero status now on shaky footing, a press conference was held. Survivor Pablo Delgado (who was studying to become a lawyer) told reporters that Christ's Last Supper had inspired him and the other survivors. Basically, Delgado explained, since Jesus had shared his own body with his disciples, it was okay that they had done the same with their deceased comrades. After hearing this explanation, even the skeptics were won over, and soon after, the Archbishop of Montevideo made it official by absolving the young men of their cannibalism-related sins.

Years later, some of the Andes survivors admitted that relating their cannibalistic acts to the sacrament was actually more of a public relations exercise than a religious experience. According to survivor Carlos Paez, "We were hungry, we were cold and we needed to live—these were the most important factors in our decision."

With this in mind, it is now time to examine the phenomenon of survival cannibalism.

# 12: The Worst Party Ever

*It is a long road and those who follow it must meet certain risks; exhaustion and disease, alkali water, and Indian arrows will take a toll. But the greatest problem is a simple one, and the chief opponent is Time. If August sees them on the Humboldt and September at the Sierra—good! Even if they are a month delayed, all may yet go well. But let it come late October, or November, and the snow-storms block the heights, when wagons are light of provisions and the oxen lean, then will come a story.*

—George R. Stewart, *Ordeal by Hunger*, 1936

It was late June, and by the time we arrived at Alder Creek, the air at snout level (which was currently about an inch off the ground) had risen to an uncomfortable 105 degrees Fahrenheit. Kayle, a five-year-old black-and-white border collie, raised her head, searching in vain for a breeze. There was a rustling in the brush nearby and something (probably a chipmunk) provided a welcome distraction to the task at hand. Kayle took a step toward the commotion.

"No," came a calm voice. It was Kayle's owner and handler, John Grebenkemper. "Go to work." Work, in this case, referred to Kayle's training as a HHRD dog, which was an abbreviation for Historical

Human Remains Detection. In short, Kayle was searching for bodies—old ones.

The dog responded instantly, reversing direction while lowering her nose to the ground. I hitched my backpack higher and followed, taking a moment to survey the meadow where Kayle slowly sniffed her way in the direction of a large pine tree. At an elevation of 5,800 feet, we were in the foothills of the Sierra Nevada Mountains, just across the Nevada border and into California. It had been a dry spring throughout the American West, and the fist-sized clumps of grass that had sprouted from the rocky soil were already turning brown. We'd passed several creek beds and I remembered reading about the muddy conditions that had led to the construction of a low boardwalk for the tourists visiting the incongruously named Donner Camp and Picnic Area.

*No need for a boardwalk today*, I thought.

We headed farther and farther away from the trail and into a mountain meadow strewn with wildflowers: orange-colored Indian paintbrush, yellow cinquefoils, purple penstemon. I'd come to the Alder Creek historic site to learn about the Donner Party, a subject I had initially planned to explore only in passing. I mean, who would be interested in yet another rehashing of what was probably the most infamous example of cannibalism in U.S. history?[16]

---

16 The tale of the Donner Party wasn't the only cannibalism-related story to emerge from the American West. In February 1874, gold prospector Alfred (or Alferd) Packer led a party of five men into Colorado's San Juan Mountains. When weather conditions deteriorated, he murdered and ate them. When the bodies were discovered the following spring, four of the five had been completely stripped of flesh. Although the skeletons showed signs of butchering, each was relatively complete and the bones showed no signs of smashing or

But as I began to investigate the Donners, I realized that research into the tragedy was alive and well, and that there were many important aspects of the story that were still unfolding.

In the summer of 1846, 87 pioneers, many of them children accompanying their parents, set out from Independence, Missouri, for the California coast, eventually taking what might qualify as the most ill-advised shortcut in the history of human travel. Dreamed up by a promoter who had never taken the route himself, the Hastings Cutoff turned out to be 125 miles longer than the established route to the West Coast. It was also a far more treacherous trek, forcing the travelers to blaze a fresh trail through the Wasatch Range before sending them on an 80-mile hike across arid lowlands that transitioned into Utah's Great Salt Desert. Tempers flared as wagons broke down and livestock were lost, or stolen, or died from exhaustion. People also died. Some from natural causes (like tuberculosis), while others were shot (by accident) and stabbed (not by accident). As the heat of summer transitioned into the dread of fall, the travelers found themselves in a desperate race to cross the Sierra Nevada Mountains before winter conditions turned the high mountain passes into impenetrable barriers. Along the way, 60-year-old businessman George Donner had been elected leader of the group, though he had no trail experience.

---

cooking. Packer had no need to process the skeletons further, presumably because he had enough meat to survive until the spring. During Packer's sentencing, the judge was rumored to have made the following statement: "There were only seven Democrats in Hinsdale County, and you ate five of them, you depraved Republican son of a bitch!"

On September 26, 1846, the wagon train finally rejoined the traditional westward route. Lansford Hastings's shortcut had delayed the Donner Party an entire month—with potentially catastrophic consequences. Disheartened, the pioneers followed the well-worn Emigrants' Trail along the Humboldt River, which by that time of year had been reduced to a series of stagnant pools. As they traveled along the Humboldt, raids by Paiute Indians further depleted their weary and emaciated livestock.

By October, any ideas of maintaining the wagon train as a cohesive unit had been abandoned. Instead, bickering, stress, exhaustion, and desperation split the group along class, ethnic, and family lines. Those travelers who could not keep up fell farther and farther behind. Afraid to overburden their oxen or slow down his own family's progress, pioneer Louis Keseberg had informed one of the older men, a Mr. Hardcoop (none of the survivors could remember his first name), that he would have to walk. Hardcoop was having an increasingly difficult time with his forced march and eventually he was left behind on the trail. Another elderly bachelor was murdered by two of the teamsters (men tasked with driving the draft animals) accompanying the group.

By the end of October, it still appeared that most of the Donner Party had overcome terrible advice, challenging terrain, short rations, injuries, and death. With the group now split in two and separated by a distance of nearly ten miles, those accompanying the lead wagons stood before the final mountain pass, three miles from the summit and a mere 50 miles from civilization. They decided to rest until the following day. But on the night before they were to make their final push, and weeks before the first winter storms usually arrived, something awful occurred.

It began to snow.

On the morning of November 1, the 59 members of the Donner Party in the lead group awoke to discover that five-foot snowdrifts had obliterated the trail ahead of them, transforming what they had expected to be a final dash through a breach in the mountains into an impossible task. It soon became apparent that there would be no crossing over the Sierras until the following spring. And so the dejected pioneers were forced to turn back, leaving behind the boulder-strewn gap that would become known as the Donner Pass.

A DAY BEFORE our trek across Alder Creek meadow, I had stood with Kristin Johnson and two of her colleagues in the Donner Pass, at the very same spot where the cross-country journey of the Donner Party had come to a halt. Looking down from the mountain, I was suddenly impressed by how resourceful and tough these pioneers had been to have made it even this far.

"I'd have never gotten up here," I said, before gesturing toward the lake that stretched to the far horizon, far below. "I would have died way down there somewhere."

Johnson, who I'd only met the day before, thought I was joking, but then the look on my face told her I wasn't.

The two of us had been corresponding about the famous pioneers for several years, and we'd finally flown into Reno; Johnson from her home in Salt Lake City and me on the Sardine Express out of JFK. After renting a car, we headed into the Sierras to meet up with Johnson's friends, former private investigator Ken Dunn and Kayle's human partner, John Grebenkemper. I'd found Johnson to be friendly, funny, and gregarious. She was also a walking, talking encyclopedia regarding anything remotely related to the Donner

Party, which was mostly fascinating but could be exasperating as well.

"I'm sure she knows what color underwear the Donners all wore," I told my wife, Janet, during a phone call later that night.

But Johnson, an enthusiastic historian and researcher, was also living proof that many of the mysteries surrounding the Donner Party remained unsolved, including the one we would be working on at Alder Creek. It was a mystery that involved the leader of the Donner Party—the very person for whom the group had been named.

On Nov 1, 1846, the pace-setting travelers whose journey had been halted at the mountain pass that would one day bear their name, decided to backtrack several miles to Truckee Lake (now Donner Lake), where they had passed an abandoned cabin that the members of a previous wagon train had constructed two years earlier. Now they would overwinter there. The pioneers quickly built two more cabins and crowded in as best they could. It was a decision that would be, perhaps, their greatest mistake.

With the benefit of hindsight, questions have arisen as to why the Donner Party did not simply backtrack another 30 miles, which would have enabled them to overwinter outside of the Sierras altogether. Among the possible explanations was their utter lack of knowledge about exactly where they had chosen to camp. Unlike other wagon trains, they had hired no seasoned mountain men to guide them.

The 21 members of the Donner Party who had lagged behind the pacesetters never made it to Truckee Lake. Nor did they experience the crushing disappointment of the final mountain pass. A

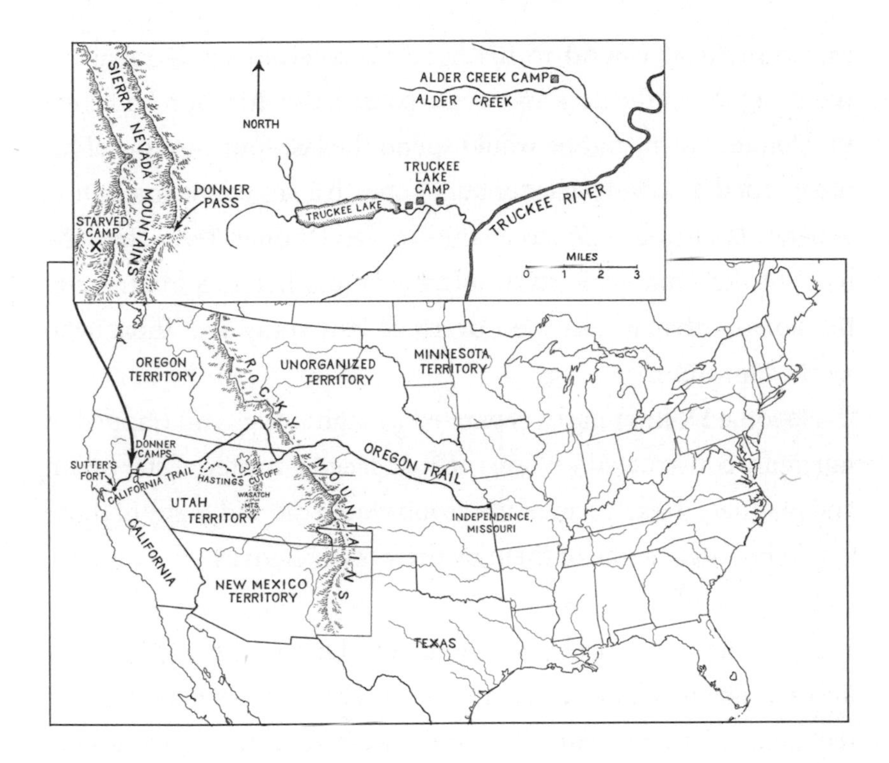


broken wagon axle had halted the group, which included George Donner, his brother Jacob, their families, and several teamsters. They eventually made it to the Alder Creek Valley, two miles west of the emigrant trail and eight miles from the Truckee Lake cabin, when the winter storm caught them completely in the open. According to survivor Virginia Reed, they "hastily put up brush sheds, covering them with pine boughs." Although the intention seems to have been to use Alder Creek as a quick rest stop before a final push into California, the weather and their weakened conditions dictated that, like those stranded at Truckee Lake, there would be no further travel until the spring thaw.

By now, George Donner had been incapacitated by what began

as a superficial wound to his hand he received while repairing their wagon. As the days and weeks passed, the infection had crept up Donner's arm, and he would spend the last four months of his life trapped in a drafty shelter built beneath a large pine that future generations would refer to as the George Donner Tree. Here the head of the Donner Party would become a helpless observer of the horrors that would soon overtake his family and those who worked for him.

Now the Donner Party, separated by eight miles and trapped in hurriedly constructed versions of Hell, faced a winter of starvation and madness. Nearly half of the group would die and many of those would be eaten, some of them by their own relatives.

THE DAY AFTER standing atop Donner Pass (where the only requirements we faced had been a reliable set of brakes in our rental car), Kristin, John, Ken, and I were at Alder Creek, hiking away from the well-worn trails and their educational signage. Kayle led us around an L-shaped stand of ponderosa pines and into a meadow covered with white flowers.

*She looks different now*, I thought, watching the dog at work. *No distractions.*

With her nose to the ground, the border collie made several passes over a bare-looking patch of ground, halting abruptly several times, only to double back over the same spot. Then she stopped, sat, and quickly pointed her nose to the ground. As my companions and I watched, Kayle stood up, moved about a yard farther, and repeated the same motion.

John turned to me. "Those are alerts. Two of them." I had learned previously that when a HHRD dog detects the scent of decomposed human remains, it responds with a trained action (like sitting) called an alert.

"But we're a half mile from the 2004 archaeological dig," I responded. "Nowhere near where the bodies are supposed to be buried."

Kayle's handler flashed a wry smile, then turned to his dog, "Good girl, Kayle. Good girl."

ON DECEMBER 16, 1846, a party of 17 men, women, and children stranded at the Truckee Lake Camp fashioned snowshoes and attempted a break-out. Early on, two of them who had started the trip without the makeshift footwear decided to turn back. The group of now 15, which also included a pair of Miwok Indians who had joined the company in Nevada, would become known as "The Forlorn Hope," and they would be making their attempt through the heart of a storm-blasted winter in the country's snowiest region.[17] According to Kristin Johnson, sometime around January 12, the survivors stumbled into a small encampment of local Indians who gave them what food they could spare (mostly seeds and acorn bread).[18] They guided the wraith-like figures partway down the mountain, but they did so warily. The pitiful travelers were not only frozen, but some of them had become seriously unhinged.

On January 17, 1847, Forlorn Hope member William Eddy reached the Johnson Ranch (no relation to Kristin Johnson), located at the

17 Alternately known in the literature as the "snowshoe group," I used "The Forlorn Hope" to avoid confusion.

18 The Nisenan (sometimes referred to as the Maidu) were the indigenous people of the Sierra Nevada foothills.

edge of a small farming community in the Sacramento Valley. By the time he staggered up to one of the cabins, Eddy looked more like a skeleton than a man. The skin of his face was drawn tightly over his skull and his eyes were sunken deeply into their sockets. His appearance sent the cabin owner's daughter away from her own front door shrieking in terror. Several horrified locals reportedly retraced William Eddy's bloody footprints into the forest and discovered six more survivors—a man and five women. The Forlorn Hope had departed the Truckee Lake Camp 33 days earlier with barely a week's worth of short rations. Eight of them eventually perished—all males, and according to Kristin Johnson, "there's no question" that seven of the dead were cannibalized.

Nearly 160 years later, science writer Sharman Apt Russell wrote about the results of a 1944–1945 Minnesota University study on the effects of semi-starvation.

> Prolonged hunger carves the body into what researchers call the asthenic build.[19] The face grows thin, with pronounced cheekbones, atrophied facial muscles account for the 'mask of famine,' a seemingly unemotional, apathetic stare . . . the clavicle looks sharp as a blade . . . Ribs are prominent. The scapula(e) . . . move like wings. The vertebral column is a line of knobs . . . the legs like sticks.

Had modern physicians been present to monitor the surviving members of The Forlorn Hope, in all likelihood these unfortunates would have exhibited most of the physiological signs of starvation: low resting metabolic rates (the amount of energy expended at rest

19 I.e., debilitated, lacking strength or vigor.

each day), slow, shallow breathing, and lower body temperatures (which would have been present even without the frigid conditions).[20] Another bodily response to starvation is low blood pressure, a condition that can lead to fainting, especially upon standing up. Like the lethargic movements that characterize starving people, these physiological changes are the body's involuntary attempts at conserving energy.

Changes in the starved body occur at the biochemical level as well, and in the case of the Donner Party, catabolic biochemical pathways would have internally mimicked the cannibalistic behavior to come, this time at the cellular and molecular level.[21] In other words, their hunger-wracked bodies would have begun to consume themselves. At first, carbohydrates stored in the liver and muscles would have been broken down into energy-rich sugars. Fat, an energy-packed connective issue (which also functions as an insulator and shock absorber in places like joints and around organs like the kidneys), would have been metabolized next. Depending on the individual, these fat stores could have lasted weeks or even months. Finally, proteins, the primary structural components of muscles and organs, would have been broken down into their chemical components, amino acids. In effect, during the latter stages of starvation, the body's system of metabolic checks and balances hijacks the energy it requires, obtaining it from the chemical bond energy that had previously been used to hold together complex protein molecules. This protein breakdown (in places like the

20 Fasting or starving people often exhibit increased sensitivity to cold.

21 Catabolic reactions (from the Greek *kata* = downward + *ballein* = to throw) are those in which larger molecules are broken down into smaller molecules, releasing energy. Anabolic reactions work in the opposite direction.

skin, bones, and skeletal muscles) produces the wasted-away look that characterizes starvation victims.

Besides physiological and behavioral effects of starvation, researchers have identified changes that occur in groups experiencing food shortages or famines. In 1980, anthropologist Robert Dirks wrote that social groups facing starvation go through three distinct behavioral phases. During the first phase, the activity of the group increases, as do "positive reciprocities." This can be thought of as an initial alarm response during which group members become more gregarious as they confront and attempt to solve the problem. Although emotions may run high, communal activity increases for a short time. The second phase occurs as the physiological effects of starvation begin to exhibit themselves. During this time, energy is conserved and the group becomes partitioned, usually along family lines. Non-relatives and even friends are often excluded. Acts of altruism decline in frequency with a concurrent increase in stealing, aggression, and random acts of violence.

The third or terminal phase of starvation is often characterized by a complete collapse of anything resembling social order. Efforts at cooperation also fall off, even within families. The rate of physical activity also decreases to near zero as the exhausted and starving individuals remain motionless for hours, basically doing nothing. Some victims of starvation do not fall into these broad patterns. These individuals are capable of heroic gestures. They are also capable of murder and cannibalism—and sometimes both.

In his book *The Cannibal Within*, Lewis Petrinovich argues convincingly that survival cannibalism is an evolved human trait that functions to optimize the chances of survival (and thus, reproductive success) for the cannibal. "It is not advantageous to be a member of another species, of a different race, or even to be a stranger when people are driven by starvation. The best thing to be is a member of a family group, and not be too young or too old."

ONLY THREE YEARS before the Minnesota University study, which came to be called the Minnesota Experiment, starvation was taking place on a massive scale in a major European city. For the inhabitants of Leningrad, the horror extended beyond the limits of a supervised research project.

Today known as St. Petersburg, Leningrad was a major industrial city and the birthplace of the Russian Revolution. In June 1941, Adolf Hitler launched Operation Barbarossa—a massive, three-pronged assault against the Soviet Union. By September, the nearly three million Leningraders were completely surrounded by German and Finnish forces. With little advance preparation by the local authorities, food shortages and dwindling fuel supplies had become grave concerns. The city's zoo animals were killed and consumed, and soon after, people began butchering and eating their pets. In a textbook example of shortsightedness, most of the city's food reserves were housed in a series of closely spaced wooden structures that were destroyed after a single bombing raid by the Luftwaffe.

On September 29, 1941, Adolf Hitler wrote, "All offers of surrender from Leningrad must be rejected. In this struggle for survival, we have no interest in keeping even a proportion of the city's population alive." German commanders were forbidden from accepting any type of surrender from the city's inhabitants. "Leningrad must die of starvation," Hitler declared.

With essential supplies all but cut off, living conditions within the embattled city plummeted along with the temperatures, which routinely reached -30 degrees Fahrenheit, in what became a winter of record-breaking cold. Although daily artillery and aerial bombardments claimed citizens at random, far more Leningraders died of exposure, sickness, and especially starvation. As a result, by December 1941, the unburied dead were accumulating by the tens of thousands.

As conditions worsened, social order began to unravel and violent criminals took to the streets. Leningrad's citizens were robbed or murdered for the food they carried home from the market or for the ration cards that allotted them as little as 75 grams of bread per day.[22]

According to historian David Glantz, 50,000 Leningraders starved to death in December 1941 and 120,000 died in January 1942. Archivist Nadezhda Cherepenina reported that during the month of February 1942, "the registry offices recorded 108,029 deaths (roughly 5 percent of the total population)—the highest figure in the entire siege."[23]

Pulitzer Prize–winning *New York Times* correspondent Harrison Salisbury wrote that once the harsh winter took hold, most of Leningrad's population was reduced to eating bark, carpenter's glue, and the leather belt drives found in motors. But there were exceptions. "These were the cannibals and their allies—fat, oily, steely-eyed, calculating, the most terrible men and women of their day."

---

22 In a system designed to maximize industrial output, Leningrad's blue-collar workers received the greatest food allowance, followed by white-collar workers, and finally dependents (who received as little as the equivalent of one and a half slices of additive-adulterated bread per day). Rations were reduced a total of five times between September and November 1941.

23 Most estimates put the eventual civilian death toll at somewhere between 800,000 and 1.5 million.

As rumors of cannibalism swept the city, so too did reports of kidnappings. It was said that children were being seized off the streets "because their flesh was so much more tender." Women were apparently a popular second choice because of the extra fat they carried.

"In the worst period of the siege," a survivor noted, "Leningrad was in the power of the cannibals."

Just as ominous, perhaps, was the sudden availability of suspicious-looking meat in Leningrad's central market. The traders were new as well, selling their grisly wares (which they claimed to be horse, dog, or cat flesh) to those shoppers with enough money to buy them. According to numerous survivor accounts, meat patties made from ground-up human flesh were being sold as early as November 1941.

Also detailed were the gruesome finds made by those assigned to deal with the thousands of dead bodies that were stacking up at the city's largest cemeteries and elsewhere. After dynamiting the frozen ground, "[the men] noticed as they piled the corpses into mass graves that pieces were missing, usually the fat thighs or arms or shoulders." The bodies of women with their breasts or buttocks cut off were found, as were severed legs with the meat cut away. In other instances, only the heads of the deceased were found. People were arrested for possessing body parts or the corpses of unrelated children.

But beyond the diaries and the accounts of Leningraders who lived through the siege, what other evidence for cannibalism has been uncovered? No physical evidence survives, no bones with cut marks suggestive of butchering or signs that they had been cooked. The inhabitants of Leningrad buried their dead, as difficult as the task had been, then tried to get beyond the nightmare they'd lived through.

As for official word, "You will look in vain in the published official histories for reports of the trade in human flesh," Salisbury wrote in 1969, and this remained so until relatively recently. All mention of cannibalism-related incidents had been purged from the public record, apparently because Stalin and other Communist Party leaders wanted to portray Leningrad's besieged citizens as heroes. Leningrad was the first of 12 Russian cities to be award the honorary title "Hero City" for the behavior of its citizens during World War II. That said, anything as unpleasant as eating one's neighbors would have cast Leningraders in something far dimmer than the glorious light mandated by their leaders.

In 2004, the official reports made right after the war by the NKVD (People's Commissariat for Internal Affairs) were released.[24]

---

24 During the Stalin era, the NKVD was a law enforcement agency, closely associated with the Soviet secret police.

They revealed that approximately 2,000 Leningraders had been arrested for cannibalism during the siege (many of them executed on the spot). In most instances, these were normal people driven by impossible conditions to commit unspeakable acts. Cut off from food and fuel and surrounded by the bodies of the dead, preserved by the arctic temperatures, Leningrad's starving citizens faced the same difficult decisions encountered by other disaster survivors, that is: Should they consume the dead or die themselves? According to an array of independent accounts as well as those from the NKVD, many of them chose to live.

ON DECEMBER 26, 1846, only ten days after leaving the Truckee Lake Camp, the members of The Forlorn Hope were lost deep in the frozen High Sierras. Only a third of the way into their nightmarish trek, they reportedly decided that without resorting to cannibalism they would all die. At first the hikers discussed eating the bodies of anyone who died, but soon they began to debate more desperate measures: drawing straws with the loser sacrificed so that the others might survive.

It was a procedure that had become known to seafarers as "the custom of the sea," a measure that provided (at least in theory) some rules for officers and their men should they find themselves cast adrift on the open ocean. Sailors drew straws, with the short straw giving up his life so that the rest might eat. In some descriptions, the person drawing the next shortest straw would act as the executioner. Although heroic in concept and theoretically fair in design, modified versions of "the custom of the sea" were sometimes less than heroic and anything but fair.

In perhaps the most famous case, in 1765, a storm demasted

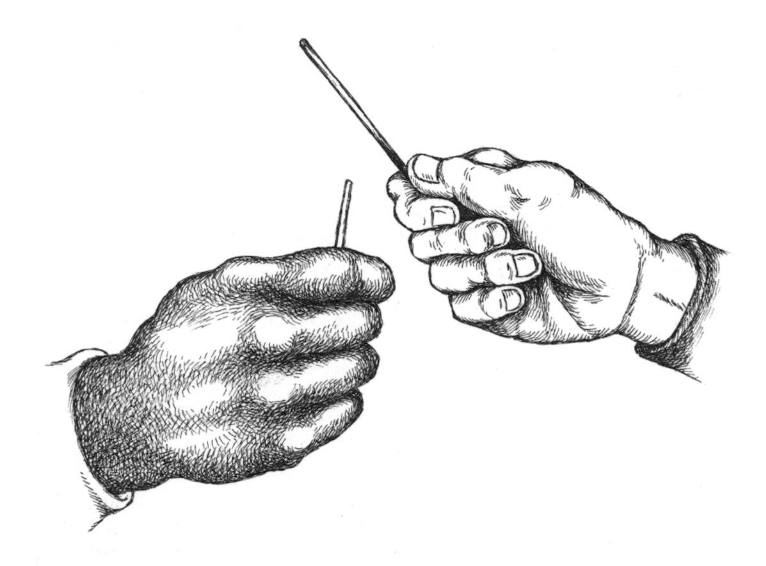

the American sloop *Peggy*, leaving her adrift in the middle of the Atlantic Ocean. On board were the captain, his crew of nine, and an African slave. They had been en route to New York from the Azores with a hold full of wine and brandy. After a month, they had nothing to eat but plenty to drink, a fact driven home when the spooked captain of a potential rescue vessel took one look at the *Peggy*'s ragged-looking crew of drunks and promptly sailed away. The *Peggy*'s captain, perhaps fearing for his own life, remained in his cabin, armed with a pistol.

Soon after the alcohol-thwarted rescue, the *Peggy*'s first mate appeared below decks, informing his captain that the men had already eaten the ship's cat, their uniform buttons, and a leather bilge pump. They had decided to draw lots, with the loser served up as dinner. The captain waved the mate away with a loaded pistol but the man returned moments later to report on the lottery results. By an incredible coincidence, the slave had drawn the short straw. Although the "poor Ethiopian" begged for his life, the captain was unable to prevent the man's murder, later writing that as they prepared to cook the body, one sailor rushed in, tore away the slave's liver, and ate it raw.

Three days later the line jumper was said to have gone insane and died. Then, in a demonstration that the crew of the *Peggy* had lost none of its well-honed survival skills, they tossed their mate's body overboard, fearful of the harmful effects of consuming a crazy man. Soon another round of straw-drawing took place, but this time the most popular and competent sailor drew the stubby stick (in this case an inked slip of paper). After making a final request that he be killed quickly, the man's drunken shipmates acted accordingly and gave him a 12-hour reprieve, during which the doomed man reportedly went deaf and lost the remainder of his mind. Just before the *Peggy*'s second homicide/buffet was set to commence, a rescue ship was spotted. Now, though, the crew feigned sobriety long enough that they were actually rescued (although they nearly forgot the evening's main course, whom they had locked below deck). In an appropriately downbeat end to the story, the reprieved man reportedly never recovered his hearing or his sanity.

LOST IN THE Sierra Nevada mountains with no food, the members of The Forlorn Hope also drew lots to determine who would be killed to provide food for the others. Patrick Dolan, a 35-year-old bachelor from Dublin, was the loser. At this point, though, no one had the heart, or possibly the strength, to carry out the killing. Someone suggested that two of the men fight it out with pistols "until one or both was slain" but this proposal was also rejected. Two days later, and before they could reconsider their options, a snowstorm rendered these choices unnecessary. Three of the group members, including Patrick Dolan, died during the night.

The next morning, according to historian Jesse Quinn Thornton, after one of The Forlorn Hope survivors was able to light a fire, "his miserable companions cut the flesh from the arms and legs of Patrick Dolan, and roasted and ate it, averting their faces from each other and weeping." Parts of the other corpses were eaten over the next few days, but it wasn't long before the survivors ran out of food again.

By now the survivors of The Forlorn Hope were exhibiting another symptom of starvation: They were bickering amongst themselves. A 30-year-old carpenter, William Foster, reportedly suggested that they kill and eat three of the women (presumably not his own wife), but when this idea failed to take hold he proposed that they shoot their Indian companions, Luis and Salvador, instead. The two men registered their votes by slipping away from the camp. Foster and the others eventually came upon them somewhere along the trail and there are several versions of what happened next.

In most accounts, Foster murdered the men, about whom little is known except that they had risked their lives on multiple occasions to save the stranded pioneers. In another version, Salvador was already dead when the hikers discovered them and Luis died an hour later. But however these men died, there is agreement on what happened next. According to John Sinclair, the *alcalde* (municipal magistrate) of Sacramento, who later presided over hearings related to the tragedy, "Being nearly out of provisions, and knowing not how far they might be from the settlements, they took their flesh likewise." Foster, who survived the whole ordeal, was never prosecuted, nor did he garner much blame for the incident. Most descriptions of the murders portray Foster's actions as being those of a decent man deranged by starvation.

Back in the mountain camps, more people were dying, and by the midpoint of The Forlorn Hope's dreadful trek, four men at the Alder Creek campsite, including George Donner's younger brother, Jacob, had perished.

Beginning in early 1847, four rescue parties (First through Fourth Relief) trekked into and out of the Sierras in fairly rapid succession. They met with varying degrees of success, tempered by cowardice, greed, and inhumanity. There was weather-related mayhem along the trail and there were deaths as well. During the ill-fated Second Relief, a blizzard forced rescuers to abandon two families of Donner Party survivors at what became known as "Starved Camp."[25] Alone on a mountain trail they thought would lead them to safety, the 13 starved pioneers huddled in a frozen snow pit for 11 days. Three of them died and the survivors were forced to eat their own dead relatives, including children. They were eventually discovered by members of Third Relief, several of whom led them out of the Sierras and to safety.

One month earlier, in mid-February, First Relief, minus several men who had decided to quit, crossed the high mountain pass where the Donner Party had been halted in November. They set up camp for the night, building their fire on a platform of logs that sat atop snow they estimated to be around 30 feet deep. The following day, seven men descended the eastern slope of the Sierras and set out across the icy expanse of Truckee Lake, arriving at the spot

---

25 Starved Camp is thought to been in Summit Valley, California, just west of Donner Pass.

where the survivors of The Forlorn Hope had told them the cabins were located. First Relief member Daniel Rhoads told historian H. H. Bancroft what happened next.

> We looked all around but no living thing except ourselves was in sight and we thought that all must have perished. We raised a loud halloo and then we saw a woman emerge from a hole in the snow. As we approached her several others made their appearance in like manner, coming out of the snow. They were gaunt with famine and I never can forget the horrible, ghastly sight they presented. The first woman spoke in a hollow voice very much agitated and said "are you men from California or do you come from heaven?"

The First Relief rescuers were shocked by the condition of the survivors. Many of the skeletal figures could barely move as they spoke in raspy whispers, begging for bread. Some appeared to have gone mad. Others were unconscious as they lay on beds made of pine boughs. The stunned Californians handed out small portions of food to each of the survivors—biscuits and beef mostly—but that night a guard was posted to ensure that their provisions would remain safe from the starving pioneers.

Outside the cabin, the members of the rescue party saw smashed animal bones and tattered pieces of hide littering the area. Then there were the human bodies, twelve of them, scattered about the campsite, some covered by quilts, others with limbs jutting out of the snow. There were no signs of cannibalism.

The next day the weather broke clear, and three of the First Relievers headed for the Alder Creek Camp. In a pair of tentlike shelters they found Tamzene Donner (George's wife), her newly widowed sister-in-law Elizabeth (who could barely walk), the twelve Donner children, and several others, including George Donner. Feverish and infirm, his wounded hand had become a slow death sentence.

Taking stock of the situation, Reason Tucker, co-leader of First Relief, knew that they needed to get out of the Sierras before another storm trapped them all there. Tucker's other realization was a difficult one, for he knew it would be physically impossible for many of the starving pioneers to hike out with them. Some were too young, others too far gone, and although he and his men had cached provisions along the trail, there would not be enough food for the entire group. It was now time for some painful decisions.

Sickly Elizabeth Donner decided that four of her children would never make it through the deep snow, and so they would remain with her at Alder Creek. George Donner's wife, Tamzene, on the other hand, was healthy enough to travel and she was urged to leave with her five daughters. Mrs. Donner refused, insisting that she would never leave George alone to die. She decided to keep her three youngest children with her, presumably waiting for the next relief party, whose arrival they apparently believed to be imminent.

On February 22, six members of the Alder Creek Camp hiked out with First Relief, accompanied by 17 others from the Truckee Lake Camp. That left 31 members of the Donner Party still trapped and starving.

A long week later, members of Second Relief arrived at the mountain camps, but by then the conditions at both sites had taken a dramatic downturn. In late 1847, reporter J. H. Merryman published the following account, obtaining his information from a letter penned by Donner Party member James Reed. Exiled earlier in the journey for stabbing a man to death in a fight, Reed had ridden on to California. Now he had returned, leading Second Relief:

> [Reed's] party immediately commenced distributing their provision among the sufferers, all of whom they found in the most deplorable condition. Among the cabins lay the fleshless bones and half-eaten bodies of the victims of famine. There lay the limbs, the skulls, and the hair of the poor beings, who had died from want, and whose flesh had preserved the lives of their surviving comrades, who, shivering in their filthy rags, and surrounded by the remains of their unholy feast looked more like demons than human beings.

In 1849, J. Q. Thornton (who also interviewed James Reed in late 1847) wrote the following about Reed's initial entry into one of the Truckee Lake cabins:

> The mutilated body of a friend, having nearly all the flesh torn away, was seen at the door—the head and face remaining entire. Half consumed limbs were seen concealed in trunks. Bones were scattered about. Human hair of different colors was seen in tufts about the fire-place.

Reed soon headed toward the Alder Creek Camp, where Thornton's account continues:

> They had consumed four bodies, and the children were sitting upon a log, with their faces stained with blood, devouring the half-roasted liver and heart of the father [Jacob Donner], unconscious of the approach of the men, of whom they took not the slightest notice even after they came up. Mrs. Jacob Donner was in a helpless

> condition, without anything whatever to eat except the body of her husband, and she declared that she would die before she would eat of this. Around the fire were hair, bones, skulls, and the fragments of half-consumed limbs.

Second Relief departed the camps on March 1, but their blizzard-interrupted trek out of the mountains would become yet another misadventure.

When the small party of men that made up Third Relief arrived at the mountain camps nearly two weeks later, they found further scenes of horror at the cabins and more dead bodies at Alder Creek. With the last of her surviving children finally accompanying the rescuers, Tamzene Donner turned down one last opportunity to save herself, deciding instead to return to the side of her frail husband. When George Donner died in late March, she wrapped his body in a sheet, said her last good-byes, and headed back to the Truckee Lake Camp. It would be Tamzene's final journey.

Donner Party member Louis Keseberg (who had not come down from the mountain because of a debilitating wound to his foot) later testified that Mrs. Donner had stumbled, half frozen, into his cabin one night. She had apparently fallen into a creek. Keseberg said that he had wrapped her in blankets, but found her dead the next morning. Sometime after the Fourth Relief (in reality a salvage team) showed up on April 17, their leader, William Fallon, wrote in his diary, "No traces of her person could be found." There was no real mystery, though, since by his own admission Keseberg, whom they had found alive, had eaten Mrs. Donner as well as many of those who had died in the mountain camps. In fact he had been eating nothing but human bodies for two months.

On April 21, 1847, Fourth Relief, accompanied by Louis Keseberg, left the Truckee Lake Camp, and four days later they reached Sutter's Fort (in current-day Sacramento). The last living member of the Donner Party had come down from the mountains.

That summer, General Stephen Kearny and his men were returning east after a brief war with Mexico. They stopped at the abandoned camp, finding "human skeletons . . . in every variety of mutilation. A more revolting and appalling spectacle I never witnessed," wrote one of Kearny's men.

The general ordered the men in his entourage to bury the dead, but instead they reportedly deposited the mostly mummified body parts in the center of a cabin before torching it. At Alder Creek, Kearny and his men found the intact and sheet-wrapped body of George Donner. There is no consensus about whether they buried him or not.

ALTHOUGH THE TALE of the Donner Party has become one of the darkest chapters in the history of the American West, time has also transformed it into something else. The dead pioneers who stare at us blankly from cracked daguerreotypes are too often a source of amusement ("Donner Party, your table is ready.") and the butt of macabre jokes. To a public that has, for the most part, become anesthetized to the concepts of gore and gruesome death, the Donner Party is no longer the stuff of nightmares. Instead, any thoughts we might have about these pioneers usually relate to vague notions about cannibalism or perhaps the perils of taking ill-advised shortcuts.

In the spring of 2010 all that changed. The long-dead travelers were back in the news, and this time the story behind the renewed media interest was neither funny nor lurid. It was actually quite remarkable. In the 1920s, schoolteacher Peter Weddell had studied the Alder Creek area and posted signs, pointing out the presumed locations of Donner Party campsites. Although he never formally presented evidence for just how he'd come to his conclusion, sites like the George Donner Tree and the Jacob Donner Camp became popular and well-marked stops at what is now the Donner Party Picnic Ground/Historical Site. Although it took more than 80 years, modern science was finally able to show that Weddell's camp localities were simply wrong.

In 2003–2004, an archaeological team from the University of Montana and Appalachian State University unearthed the remains of a campsite at Alder Creek that would become known as the Meadow Hearth. It contained artifacts like cooking utensils, fragments of pottery, and percussion caps—small, explosive-filled cylinders of copper or brass that allowed muzzle-loaders to fire in any weather. Each of these items dated to the 1840s. There were also thousands of bone fragments, and given the Donner Party's reputation, interest soon centered on whether or not any of these bones were human in origin.

Six years later in 2010, the researchers had completed their analysis of the artifacts and were preparing a scientific paper that would detail their findings. Now, though, and before their paper could be published, a spate of articles and news blurbs announced that the scientists had uncovered physical evidence that led them to seriously question the very act for which the Donners had attained their infamy.

"Analysis Finally Clears Donner Party of Rumored Cannibalism," read one media report, while Discovery News informed its readers that the "Donners Ate Family Dog, Maybe Not People." The original subtitle, "Did ethnic prejudice spur the now infamous legend of the Donner Party's cannibalism?" hinted at an intriguing new explanation for the notorious and long-held accusations. The subtitle resulted from the author's mistaken belief that Louis Keseberg, the most notorious Donner Party member, was Polish (he was German) and that a prejudice against Poles had led to claims that he became a subhuman cannibal when the going got tough.[26] Even the *New York Times* got into the act. "No Cannibalism Among the Donner Party?" read the bet-hedging headline in a *Times*-associated blog. My personal favorite was a headline from a blog post at *The Rat:* "Scientists Crash Donner Party."

So how did this information come about? And was there any truth to it?

Initially, the archaeological team working at Alder Creek uncovered a thin layer of ash that they eventually determined to be the remains of an 1840s-era campsite. What became known as the Meadow Hearth dig also revealed concentrations of charred wood and deposits of burned and calcined bone fragments. The latter occurs when bone is subjected to high temperatures, resulting in the loss of organic material, like the protein collagen. What's left is a mineralized version of the original bone and, importantly, one that is more resistant to decomposition than it was in its original form. Calcined bone also provides anthropologists with strong evidence that the bones in question were cooked.

---

26 The subtitle was subsequently eliminated from the online version of the article after complaints by readers.

All told, the university researchers collected a total of 16,204 bone fragments from the Meadow Hearth excavation, a number that makes it far easier to understand why it took them six years to analyze their data. Unfortunately, not everyone was as patient as the scientists had been.

On April 15, 2010, the Office of Public Affairs at Appalachian State University (ASU) issued a press release titled "Appalachian Professor's Research Finds No Evidence of Cannibalism at Donner Party Campsite." Posted on ASU's University News site, it began with the following statement:

> Research conducted by Dr. Gwen Robbins, an assistant professor of biological anthropology at Appalachian State University, finds there is no evidence of cannibalism among the 84 members of the Donner Party who were trapped by a snowstorm in the Sierra Nevada Mountains in the mid-1840s.

The piece mentioned Robbins's preliminary results and how the osteologist "had been asked to determine whether or not the bone fragments were human." Robbins, they said, analyzed 30 bone bits as a grad student and 55 more several years later while working at ASU. After using an array of histological techniques, she concluded that the bones had come from cattle, deer, horse [probably mule], and dog, but that none of the fragments could be identified as human in origin.

Next, the ASU blurb writers trotted out their big gun—statistics:

> A power analysis indicated that, statistically, Robbins and [fellow researcher] Gray can be 70 percent confident that if cannibalism

> made up a small fraction of the diet (less than 1 percent) at the site in the last few weeks of occupation, and if humans were processed in the same way animals were processed, at least one of the 85 bone fragments examined would be human.

With statistics firmly on their side, the PR scribes at ASU loaded up, took careful aim at their own feet, and fired off this bold statement:

> The legend of the Donner party was primarily created by print journalists, who embellished the tales based on their own Victorian macabre sensibilities and their desire to sell more newspapers.

They went on to add, "The survivors fiercely denied allegations of cannibalism," a statement contradicted by Donner Party survivors, rescuers, and historians alike. Finally, and as if to further convince the world that Donner Party members were actually humans and not crazed cannibals, the ASU PR crew announced that pieces of writing slate and broken china found near the cooking hearth "suggest an attempt to maintain a sense of a 'normal life,' a family intent on maintaining a routine of lessons, to preserve the dignified manners from another time and place, a refusal to accept the harsh reality of the moment, and a hope that the future was coming."

The response was predictable. Media types, from obscure bloggers to major newspaper reporters and popular science writers, latched onto the story and within days their readers were being informed that serious and scientifically based doubts had risen over the question of cannibalism by the members of the Donner Party. Formerly a textbook example of survival cannibalism, the claims of people-eating by the starving Donner Party were now being blamed on Victorian-era journalists and ethnic prejudice.

In reality, though, there was no controversy at all, at least among most Donner Party experts. The PR department at ASU had simply blown it by badly misrepresenting the study's preliminary results. The key statement by the PR mavens, and one that should have prevented the entire mess, can be found in the previously cited quotation about the statistical probability of finding a human bone among those examined by researchers. Such a discovery, they wrote, would have been statistically probable "if humans were processed in the same way animals were processed." As I've mentioned, this is a requirement for determining whether cannibalism has occurred or not, but therein lies the problem. As it turns out the Donner Party did not process human bones and animal bones in the same way, and there's a good reason why.

Of the thousands of bone fragments from the Meadow Hearth examined by researchers, 362 of them showed evidence of human processing. About one quarter of those had abrasions and scratch marks, which indicate that the bones had been smashed into bits. Other pieces of bone exhibited a condition known as "pot polish," a smoothing of the edges that results from the bones being stirred in a pot. To anthropologists this was another strong indicator that the bone fragments had been cooked.

As starvation set in, the stranded members of the Donner Party ate whatever they could find. According to historical accounts, they consumed rodents, leather belts and laces, tree bark, and a gooey pulp scraped from boiled animal hides. By the end of January 1847, they began consuming their pet dogs. The analysis by Gwen Robbins and her coworkers indicated that bones from several types of mammals had been smashed, boiled, and burned by someone at the Alder Creek Camp. This would have been done in an effort to render the bones edible, while extracting every bit of nutrient possible. In all likelihood, these would have been the types of last-resort measures undertaken before the survivors turned to cannibalism, which did not begin in the mountain camps until the last week of February 1847—sometime after the departure of First Relief on February 22 and before the arrival of Second Relief a week later. The practice of consuming dead bodies continued until the survivors either died or were rescued, and for everyone except the soon-to-be-christened Donner Party monster, Louis Keseberg, cannibalism would have lasted only a week or two at most, a vitally important point.

Given the large number of bodies present at the Truckee Lake and Alder Creek campsites, and the short amount of time during which cannibalism occurred, there would have been no need to process human bones in the same manner in which animal bones had been processed previously. Essentially, that's because once cannibalism began at the camps there would have been ample human flesh for the ever-dwindling number of survivors to eat—more than enough to make cooking and re-cooking the human bones completely unnecessary. For similar reasons, once Louis Keseberg was the only person left alive at either camp, between the human bodies and the livestock carcasses uncovered by the melting snow, he would have had plenty of food (grisly though it was) on which to subsist until his rescue by Fourth Relief a month later.

Because uncooked bones would not have been preserved in the acidic soil of the conifer-dense Sierras, there would be no human bones for archaeologists to uncover. Therefore, the absence of calcined human bones from the Meadow Hearth only proves that human and animal bodies were not processed in the same way. The evidence does not place the practice of cannibalism by members of the Alder Creek Camp into doubt, nor does it have any bearing whatsoever on the cannibalism that took place at the Truckee Lake Camp, within The Forlorn Hope, or at the Second Relief's Starved Camp.

The good news was that someone at Appalachian State yanked the press release and deactivated the global web link. A week later, ASU's University News website featured a revised version of the original release, minus the grandiose claims of its predecessor. The newly titled report, "Professor's Research Demonstrates Starvation Diet at the Donner Party's Alder Creek Camp" was a far more straightforward story on the preliminary findings of the research team. Those who had written the revision also chose not to address the previous boo-boo.

The actual research paper (published three months later in the archaeological journal *American Antiquity*) turned out to be a fine piece of science. Yet it is likely that very few people outside the archaeology/anthropology communities will ever read it, and this presents its own problem.

According to Kristin Johnson, since 2010 the claims of "no cannibalism" have made it that much more difficult for Donner Party historians to present a factual representation of the events that took place. "Unfortunately, people's memories seem remarkably retentive when it comes to misinformation," Johnson told me. "And once a falsehood or garbled story gets out, it's difficult to dislodge. 'We found no evidence for cannibalism at Alder Creek' becomes 'There was no cannibalism in the Donner Party.'"

Even the media seemed less excited about the prospect of returning to the long-held Donners-as-cannibals stance, although the *New York Post* scored points with their headline, "'Cannibal' Doc Eats Her Words."

ALTHOUGH THE 2010 media flap had certainly sparked my interest in writing about the Donner Party, I had come to Alder Creek in 2014 for a completely different reason. In brief, historians like Kristin Johnson had begun to doubt the claims of some of the 2003–2004 archaeological team that the Meadow Hearth had once been the camp of Donner Party leader George Donner, whose body had been discovered, still wrapped in its sheet, by General Kearny and his men in the summer of 1847.

"That just didn't work for me," Johnson told me, and I asked her why.

"The site didn't really fit the sources. There were no tree remains nearby. There weren't all that many artifacts [excluding bone fragments] and few indicating a female presence. Initially, I thought the dig site was more likely Jacob Donner's camp, since there were seven males and two females in his family."

But that conclusion didn't sit well with her, either.

Johnson began working with Kayle's owner John Grebenkemper. The two Donner Party detectives examined old photos of the tree stump locations and shared relevant documents including maps, articles, memoirs, and letters written by Donner Party members. Grebenkemper, a retired computer whiz with a Ph.D. from Stanford, wrote a program to coordinate historic photos with the current topography of Alder Creek. Once their analysis of the data was complete, they arranged for the area to be examined by a pack of grave-sniffing pooches, including Kayle.

In addition to George Donner, 34 members of the Donner Party died in the winter camps or trying to escape them—mostly from starvation and/or exposure. In 1990, anthropologist Donald Grayson conducted a demographic assessment of the Donner Party deaths and came up with some interesting information.

On the not-so-surprising front was the fact that children between 1 and 5 years of age and older people (above the age of 49), experienced high mortality rates (62.5 percent and 100 percent, respectively), primarily because both groups are more susceptible to hypothermia.

What I found fascinating was that 53.1 percent of males (a total of 25) perished while only 29.4 percent of females died (10). Additionally, not only did more of the Donner men die, they died sooner. Fourteen men died in between December 1846 and the end of January 1847, while females didn't begin dying until February.

Another intriguing detail is that all 11 Donner Party bachelors (over 18 years of age) who became trapped in the Sierras died, while only 4 of the 8 married men, traveling with their families, perished during the ordeal.

The explanation for why more Donner Party males died than females is probably a combination of biology and behavior.

The biological component relates to the physiological differences between males and females, and nutritional researchers believe that three significant differences come into play during starvation conditions: 1) Females metabolize protein more slowly than males (i.e., they don't burn up their nutrients as quickly as males); 2) Female daily caloric requirements are less (i.e., they don't need as much food as men); and 3) Females have greater fat reserves than males, thus they have more stored energy that can be metabolized during starvation conditions. Also, much of this fat is located just below the skin (i.e., subcutaneous), where it functions as a layer of insulation, helping maintain the body's core temperature during conditions of extreme cold.

The behavioral component of the female/male survival differential relates to the fact that the Donner Party men did most of the hard physical labor associated with a journey by wagon train, and that ultimately translated to serious health problems once their diets became compromised. Donald Grayson suggested a scenario that triggered the decline in the previously healthy males.

> When the Donner Party hacked a trail through the Wasatch Range . . . it was the men, not the women who bore the brunt of the labor. . . . There is no way to know exactly how much this grueling labor affected the strength of the Donner Party men, but they surely emerged from the Wasatch Range with their internal energy stores drained, stores they were unable to renew during the long and arduous trip across the Great Basin Desert that followed.

So what about the fact that married men out-survived bachelors by such a wide margin? The reason for this may have to do with differences in the mammalian physiological response to stress, related to blood levels of the hormone cortisol (hydrocortisone), a steroid hormone released by the adrenal gland. Cortisol is considered a stress hormone and part of the body's "fight or flight" response to real or imagined threats. While it can have positive short-term effects, increased plasma levels of cortisol can also lead to decreased cognitive ability, depression of the immune system, and impairment of the body's ability to heal.[27] In a 2010 study, researchers at the University of Chicago looked at hormone levels in test groups composed of married and unmarried college students who were placed in anxiety-filled situations. The bachelors had higher levels of cortisol than did married men subjected to the same levels of stress. Thus the experimenters concluded that

> single and unpaired individuals are more responsive to psychological stress than married individuals, a finding consistent with a growing body of evidence showing that marriage and social support can buffer against stress.

If one adds these findings to the data from Robert Dirks's study (in which one phase of starvation was for groups to partition along family lines), the results strongly indicate why all of the mountain-stranded bachelors perished while fully half of their married counterparts survived.

BACK AT ALDER Creek, while my new friend Kayle and I rested in the shade of a large pine tree, Johnson and Grebenkemper outlined their new hypothesis. As I thought, it concerned the location of what they now believed to be multiple campsites at Alder Creek, one each for the two Donner brothers, George and Jacob, and another for the teamsters who worked for them. Grebenkemper

---

27 The short-term, positive effects of cortisol release include a burst of energy (through an increase in blood sugar levels) and a lower sensitivity to pain (by reducing inflammation).

told me that in 2011 and 2012, Kayle and several other HHRD dogs had alerted at the Meadow Hearth as well as another spot (Canine Two Locality) some 500 meters from where Kayle had just alerted. I was only mildly disappointed to learn that the site near where I now sat had already been designated Canine Three Locality. These localities were well outside the area traditionally associated with Donner camps, and this particular site had been reported to have the strongest scent signature—a determination made at the end of 2013 after six different HHRD dogs alerted here a total of 27 times.

According to Grebenkemper, eight members of the Donner Party died at Alder Creek, their bodies placed in the snow, not only because of the weather conditions but because the survivors were too weak to bury them. In the spring of 1847, the thawed-out bodies began to decompose and the scent from the bones and body fluids would remain in the ground for nearly 170 years. Having gone through specialized training programs that began when they were puppies, HHRD dogs like Kayle were able to detect that scent.

Grebenkemper and Johnson believed that the Canine Localities Two and Three matched up extremely well with a combination of old maps and survivor accounts and were the best possible fit for the sites of the two Donner family campsites.

"I've had a change of mind about the teamsters' shelter," Kristin Johnson told me. "Previously, I'd cast doubts on its existence, arguing that there was no good source for it. But with this new site agreeing so well with the sources, I think that for now the best explanation for the Meadow Hearth is that it was the site of the teamsters' hut."

I gestured to the patch of dry ground in front of me where Kayle had just alerted. "So that would make that spot, over there . . . ?"

"The spot where Tamzene Donner placed George Donner's body," Grebenkemper said, nonchalantly.

I sat up straight.

He continued, quiet and calm. "And it would make that tree you're sitting under the *real* George Donner Tree."

The Donner detectives smiled wry smiles, sensing my momentary confusion as I scrambled to my feet. "Wait. And what happened under this tree?"

"A lot of suffering," John replied.

Kristin Johnson finished up. "So when and if this new site is excavated, we may have to modify our thinking about all sorts of things related to the Alder Creek camps"—most importantly, where they were actually located.

Minutes later we were hiking out of the 19th century and back to the parking area. Letting my gaze fall on the clusters of tiny white flowers that covered the meadow, I couldn't help thinking about another blanket of white that had vexed the Donner Party, perhaps at this very spot, during the long and horrible winter of 1846–1847. Though the air temperature had risen even higher since our arrival that morning, I shivered at the thought. Then I turned my face toward the warm, late-June sun and headed for my car.

WE'VE ALREADY LEARNED that cannibalism occurs across the entire animal kingdom, albeit more frequently in some groups than others. When the behavior does happen, it happens for reasons that make perfect sense from an evolutionary standpoint: reducing competition, as a component of sexual behavior, or an aspect of parental care.

Cannibalism in nature is also widely seen as a natural response to stresses like overcrowding and food shortages. The unfortunates involved in shipwrecks, strandings, and sieges have also resorted to cannibalism, and by doing so they exhibited biologically and behaviorally predictable responses to specific forms of extreme stress. Although the conditions may have been unnatural, the actions that resulted were not.

Additionally, like male spiders that give up their lives and bodies to their mates, ultimately increasing the survival potential of their offspring, so too did the bodies of Donner Party members like Jacob Donner serve a similar function for their families.

Finally, in cannibalism-related tragedies like the Donner Party, survivors have been given something like a free pass for committing acts that would otherwise be considered unforgivable by their cultures.

But where did this taboo come from? Why is the very idea of human cannibalism so abhorrent that except in a very few cases it justified the torture, murder, and enslavement of those accused of being cannibals?

# 13: Eating People Is Bad

*Baby, baby, naughty baby,*
*Hush you squalling thing, I say.*
*Peace this moment, peace or maybe,*
*Bonaparte will pass this way.*

*And he'll beat you, beat you, beat you,*
*And he'll beat you all to pap,*
*And he'll eat you, eat you, eat you,*
*Every morsel snap, snap, snap.*

—*The Oxford Dictionary of Nursery Rhymes*

The word *taboo* has a Polynesian origin, and the English explorer and navigator Captain James Cook reported that its use by the South Sea islanders related to the prohibition of an array of behaviors—from eating certain foods to coming into physical contact with tribal leaders. Unfortunately for Cook, the first official link between the terms *taboo* and *cannibalism* may have been based on his crew's initial, though evidently mistaken, fear that Cook himself had been cannibalized.

On February 14, 1779, after what turned out to be a serious misunderstanding, Cook was clubbed to death by Hawaiian islanders, who then cooked and deboned his body before divvying it up among local chiefs as a way as of incorporating him into their aristocracy. Since it was only right that Cook's own people got their share of the body, a charred section of it was returned to Lieutenant James King, who asked the Hawaiians if they had eaten the rest of it. According to King, "They immediately shewed [*sic*] as much horror at the idea, as any European would have done; and asked, very naturally, if that was the custom among us?" So while the islanders had murdered, cooked, and filleted the explorer, they *hadn't* eaten him, though the latter point is often misrepresented in accounts of the incident.

Reay Tannahill, a British historian who wrote both fiction and nonfiction, was perhaps best known for her books *Food in History* and *Sex in History*. In 1975 she wrote *Flesh and Blood*, the first scholarly study of cannibalism accessible to the general public. In it, Tannahill proposed that Judeo-Christian customs related to the treatment of the dead contributed to the strongly held belief that eating people was wrong. Specifically, she referred to the "belief that a man needed his body after death, so that his soul might be reunited with it on Judgment Day." Since cannibalism involved dismemberment as well as other procedures familiar to those in the butchery profession, it was no surprise that these practices induced in Christians and Jews alike "an unprecedented and almost pathological horror of cannibalism."

Decades later, others, like journalist and author Maggie Kilgore, addressed questions related to the prevalence of cannibalism taboos. They suggested that in addition to wanting the bodies of the dead to stick around intact until Judgment Day, our picky rituals concerning what foods could or couldn't be eaten (e.g., the Jewish ban on eating pork) were just as important.

To Kilgore, the term "you are what you eat" is a reflection of the importance of food as a "symbolic system used to define personal, national and even sexual differences." Outsiders and foreigners, according to Kilgore, are often "defined in terms of how and, especially, what they eat, and denounced on the grounds that they either have disgusting table manners or eat disgusting things." For example, the derogatory term "frogs" for French people is based on their consumption of frogs' legs—something the British (who coined the term) would presumably never do. Likewise, and once again echoing anthropologist Bill Arens's stance, calling someone a cannibal becomes a means of using dietary practices (whether real or imagined) to define a particular culture as savage or primitive.

Of course this idea leads to the question of whether cannibalism might be more frequent or more readily acceptable in cultures that don't hold Judeo-Christian beliefs about the afterlife or whose adherents follow diets with fewer religious or culturally imposed restrictions. First, though, let's investigate how the Western cannibalism taboo became so widespread—a phenomenon that began with an ancient civilization whose early writings would go on to influence both Christian and Semitic cultures.

IN ALL LIKELIHOOD, the first mention of something approaching cannibalism in Western literature occurs in Homer's epic poem *The Odyssey*, which dates from approximately the 8th century BCE. On an island stopover, the adventurer Odysseus (known as Ulysses in Latin) and his men enter the cave of Polyphemus, a Cyclops (albeit one with a human shape). Luckily, the giant is out tending his flock, so the Greeks make themselves at home: lighting a fire, eating some of the big guy's cheese, and trying to decide what else they can steal. The party ends abruptly when Polyphemus returns home and blocks their exit with an enormous stone. Odysseus tries to bluff his way out, bragging about the city he had recently sacked and presumably his soon-to-be-famous wooden horse trick. He also tells Polyphemus to be extremely careful, since he and his pals are well protected by the gods. The Cyclops, however, appears somewhat less than impressed. According to Odysseus:

> Lurching up, he lunged out with his hands towards my men and snatching two at once, rapping them on the ground he knocked them dead like pups—their brains gushed out all over, soaked the floor—and ripping them limb from limb to fix his meal he bolted them down like a mountain-lion, left no scrap, devoured entrails, flesh and bones, marrow and all!

After washing down the gruesome meal with milk, the giant falls asleep. The next day, Polyphemus consumes two more of the Greeks for breakfast and another pair for supper, and although Odysseus feels that the jury is still out on his intimidation ploy, his men suggest that he come up with an alternate plan. Soon after, our hero talks the Cyclops into drinking some wine he and his men had brought, telling the giant that they'd intended to present it to him as a gift—before he started eating everybody, that is. After downing three bowlfuls, Polyphemus falls down drunk, "as wine came spurting, flooding up from his gullet with chunks of human flesh."

Presumably skirting bits of their partially digested crewmates, the vengeance-minded Greeks uncover an oar-sized piece of wood they had previously sharpened and buried under the sheep dung littering the cave floor. After heating the tip, Odysseus and four mates snap into battering-ram mode, slamming the point home and poking out the snoozing Cyclops's favorite eye. The crafty Greeks avoid the enraged Polyphemus and also manage to make him look bad in front of his giant friends, who have stopped by to investigate the ruckus. The following morning, after the Cyclops rolls away the stone to let out his flock, Odysseus and his men make their escape—hanging beneath the bodies of the giant's sheep.

In *Theogony*, Homer's fellow poet Hesiod recounts the tale of Cronos, the Father of the Gods, who learns from his parents (Heaven and Earth) that his own son will one day overthrow him. To prevent this, Cronos eats his first four children, but the youngest, Zeus, is spared when the children's mother hands her husband a rock wrapped in swaddling clothes instead of baby Zeus.

I interviewed classicist Mary Knight at her office in the American Museum of Natural History. "The tale of Cronos suggests an early religious connection with the taboo on eating people," she said, "since Zeus would not do to his offspring what his father tried to do to him." Echoing what Maggie Kilgore wrote about how eating can be used as a way to reinforce cultural differences, Knight continued. "The story may thus support cannibalism as a part of the ancient Greek view of a 'primitive' past vs. the 'civilized' present. Greeks came to see themselves as different, calling all non-Greeks 'savages'—people who may have continued eating people."

Although Polyphemus and Cronos were fictional characters, and not exactly card-carrying humans (which might upset cannibalism purists), this may not have been the case with some of the man-eaters described by another Ancient Greek, Herodotus (ca. 484–425 BCE). In his masterpiece, *The Histories*, the man often referred to as the

Father of History wrote that the Persian king Darius asked some Greeks what it would take for them to eat their dead fathers. "No price in the world," they cried (presumably in unison). Next, Darius summoned several Callatians, who lived in India and "who eat their dead fathers." Darius asked them what price would make them burn their dead fathers upon a pyre, the preferred funerary method of the Greeks. "Don't mention such horrors!" they shouted.

Herodotus (writing as Darius) then demonstrated a degree of understanding that would have made modern anthropologists proud. "These are matters of settled custom," he wrote, before paraphrasing the lyric poet Pindar, "And custom is King of all." In other words, society defines what is right and what is wrong.

But while some of Herodotus's writings certainly reinforced the idea that "Culture is King," at the same time his tales also portrayed cannibalism as a sensational and utterly repugnant act, thus helping to propagate a mindset that cannibalism was bad behavior. As such, his combination of history and myth offers important clues about the spread of the cannibal taboo.

Herodotus was also the first writer to document the practice of drawing lots during crises, with the person holding the short straw killed and eaten by his starving comrades. According to the historian, during King Cambyses's expedition to Ethiopia, his men ran out of provisions, and after slaughtering and consuming their pack animals, they were reduced to munching on grass. Herodotus describes how when they came to the desert, "some of them did something dreadful." They cast lots with one out of ten men killed and eaten by his comrades. After learning of this, Cambyses reportedly abandoned the campaign.

The Father of History also wrote extensively about the Scythians, horse-riding barbarian nomads living in the area north of the Black Sea. According to Herodotus, among their many strange customs, the Scythians enjoyed smoking marijuana and eating their enemies. Additionally, like Ed Gein, the model for the fictional characters Norman Bates and Buffalo Bill (*The Silence of the Lambs*), Scythian warriors also found some unique uses for human skin and body parts, using severed hands for arrow quivers and carrying around human skins stretched upon frames.

In what may be Herodotus's most influential cannibal-related story, he recounted the tale of Astyages, the last king of the Median Empire. One night, the king awakens from a particularly bad nightmare in which his daughter Mandane "[made] water so greatly that she filled all his city," eventually flooding all of Asia. Several years later, as Mandane is carrying her first child, the king has another bad dream. In this one, an enormous vine grows out of "his daughter's privy parts" until all of previously-urine-soaked Asia falls under its mighty shade. The Magi are asked to interpret and they attempt to put their king at ease by telling him that Mandane will give birth to a son and that the boy will one day destroy Astyages's empire.

Since there are apparently no pruning shears big enough for this gardening job, Astyages sends his favorite general, Harpagus, to find Mandane and kill her child. Harpagus, however, refuses to spill innocent blood and instead hands the newborn off to a herdsman and his wife—the latter (by coincidence) has just given birth to a stillborn son. Predictably, the quick-thinking general departs with the body of the dead child, which he delivers to the king.

Ten years later, Mandane's son and his sheep-herding foster dad are granted an audience with King Astyages, who while talking to the boy recognizes the family resemblance. After some quick back-calculations, the king realizes what his formerly favorite general has done. Astyages sends the boy off with servants, then questions the herdsman, who quickly fesses up to the entire ruse. Harpagus is summoned, and seeing the herdsman, he attempts to weasel out of the predicament, admitting that he felt bad about killing the boy. Harpagus then tells the king that he did what anyone in his predicament would have done—he ordered the herdsman to murder the infant.

King Astyages, who may have also been famous for his poker face, then tells Harpagus something along the lines of, "Hey, no problem, I felt bad about asking you to kill my grandson anyway." The general presumably lets out a huge sigh of relief, but before he can get too relaxed, the King follows up. "Oh, and by the way," he adds (although, once again, probably not in those exact words), "why don't you and your son come to dinner tonight so we can all celebrate together?" Relieved, Harpagus returns home and instructs his son to head over to the banquet immediately. The boy responds with the ancient Persian equivalent of "You got it, Dad," and leaves for the party.

According to Herodotus, this is what happened next:

> When Harpagus's son came to Astyages, the king cut his throat and chopped him limb from limb, and some of him he roasted and some he stewed. . . . When it was dinner hour and the other guests had come, then for those other guests and for Astyages himself there were set tables full of mutton, but, before Harpages, the flesh of his own son, all save for the head and extremities of the hands and feet; these were kept separate, covered up in a basket.

When the meal is done, Astyages asks the general how he liked the feast and although Harpagus initially gives it the big thumbs up, the party ends on a sour note once the king has his general open the basket containing his son's uneaten body parts.

IF THIS STORY sounds familiar, that's because it has appeared in several versions since the time of Herodotus. Most notably, William Shakespeare co-opted it for the filial cannibalism scene in *The Tragedy of Titus Andronicus*. In the Bard's most violent and arguably most maligned play, Titus, a Roman general, engages in an increasingly gory running battle with his archenemy, Tamora, the queen of the Goths. Late in the play, and after his daughter has been raped and mutilated by Tamora's two sons, Titus exacts his revenge. He kills the siblings and has their bodies baked in a pie, which he serves at a banquet to the queen and her husband, Saturninus. After Titus reveals his secret ingredient, everyone's plans for a quiet meal get tweaked a bit when Titus kills Tamora, Saturninus kills Titus, and Titus's son kills Saturninus.[28]

It is also possible that Shakespeare may have gotten his cannibal inspiration from Seneca's 1st century Roman tragedy, *Thyestes*, in which the title character not only tricks his twin brother, Atreus, out of the throne of Mycenea, but also takes his sister-in-law as a lover. Thyestes continues his bad behavior by chiding Atreus that he can have the throne back as soon as the sun moves backward in the sky. Zeus however, overhears the taunt and "drives the day

---

28 An alternative source for Shakespeare's cannibal scene may have been the Roman poet Ovid (ca. 43 BCE–18 CE) who also lifted Herodotus's story of Astyages for parts of his own lyric poem, *The Metamorphosis*.

back against its dawning." Before you can say "banished," Thyestes is forced to surrender the throne. Atreus, though, isn't done with his slimy sibling, and after learning of his wife's infidelity, he invites Thyestes to a reconciliatory banquet. As part of the party prep, Atreus murders Thyestes's two sons from the forbidden relationship and serves them to their unsuspecting dad (who has obviously *not* been keeping up with his readings of Herodotus). At dinner's end, Atreus presents Thyestes with the hands and heads of his slain children on a platter, forever defining the term *Thyestian Feast* as one at which human flesh is served.

In short, from the Ancient Greeks to William Shakespeare, and in stories written across a span of 2,500 years, cannibalism was depicted as either the ultimate act of revenge or the gruesome work of gods, monsters, and savages (a.k.a. non-Christians and anyone living in the vicinity of some gold). By the 17th and 18th centuries, with the taboo firmly established, the threat of cannibalism would reach a new audience and serve a new purpose—as a way to terrorize children into behaving.

Jakob and Wilhelm Grimm (born in 1785 and 1786 respectively) were German academics who collected oral folk tales during the early 1800s. They did so by interviewing peasants, servants, middle-class types, and aristocrats, and they published hundreds of fairy tales in the years between 1812 and 1818. In the parade of new editions that followed, the brothers changed, added, and subtracted stories, depending on how well they had been received previously. Like the ancient Greek and Roman myths, the original fairy tales depicted violence, desire, heartbreak, and fear. They also portrayed the all-too-common hardships of their own time, especially famine and the abandonment of children by destitute parents. The language was often scatological and, as such, many of the updates the authors initiated reflected the fact that the originals were definitely not kid-friendly.

As the Grimms sanitized these tales for publication, and for a much younger readership, themes were also modified. But rather than molding them into the bedtime stories familiar to modern readers, the brothers transformed them into cautionary tales, many of which ended badly for children who chose not to obey their parents. On one level at least, fairy tales can be seen as literary relics from a time when terror was an accepted educational tool. Bearing almost no resemblance to the politically correct stories written today for kids, the original Grimm's fairy tales were tools employed by parents to socialize children, to increase their moral standing, and to frighten them into obeying the directives of their elders.

The Grimm brothers were preceded as writers by Charles Perrault (1628–1703), a Frenchman whose 1697 *Histoires ou Contes du Temps passé*, provided readers with what may have been the earliest written collection of fairy tales. His most famous book, subtitled *Les Contes de ma Mère l'Oye* (*Tales of Mother Goose*) contained eight stories, including *Red Riding Hood*, *Sleeping Beauty*, and *Puss in Boots*, and its reception by the public elevated the fairy tale into a new literary genre. As it would be with the Grimm stories, Perrault's fairy tales often contained a heavy dose of cannibalism. For example, most children and adults will recall that the wicked queen in *Snow White* wanted the title character killed. Less familiar, perhaps, is that in the original tale, the queen not only orders a huntsman to murder Snow White but to return with her liver and lungs as proof that the deed had been done. Taking pity on the innocent beauty (Harpagus style), the hunter slays a boar instead and brings the queen a Snow White–sized portion of porcine organ meat. Then, in a scene that somehow wound up on the cutting room floor at the Disney studios, the misled monarch cooks up the offal in a stew, which she eats, thinking perhaps that except for an unfortunate gravy stain, she has seen the last of Snow White.

An equally disturbing revelation is found in the source material for the Perrault fairy tale *Little Red Riding Hood.* Unaltered from Perrault's story is the setup, in which the wolf gets to Granny's house before Red. But in the original story (a French peasant tale that may date from the 10th century) as translated by Paul Larue and reported by fairy tale scholar Jack Zipes, instead of gobbling down the old woman whole (so that she can later emerge, Jonah-like, from the wolf's bisected belly), the werewolf murders the old woman and cuts her up—storing pieces of Granny meat in the cupboard, along with a bottle of her blood. When Red Riding Hood arrives, the creature directs her to the cabinet, saying, "Take some of the meat which is inside and the bottle of wine on the shelf." After unknowingly eating her own grandmother and drinking her blood, Red strips and the wolf tosses her clothes into the fire ("You won't be needing those anymore," he tells her). She then gets into bed with the hirsute granny, and after a famous bit of dialogue, Red escapes after convincing the creature that she needs to go outside for a pee (I'm not making this up).

In Perrault's *Hop o' My Thumb*, seven young brothers, led by Little Thumb, the smallest but smartest sibling, are abandoned in the forest by their destitute parents in a time of great famine. A kindly woman, who turns out to be the wife of a "cruel Ogre who eats little children," eventually takes in the lost kiddies. In the nick of time, she hides them under a bed as her giant husband returns (luckily he knocks before entering his own house), but soon he smells "fresh meat" and drags the children out from their hiding place. Even as the kids fall to their knees, begging for mercy, the ogre is already "devouring them in his mind," especially since "they would be delicate eating, when [my wife] made a good sauce."

The story ends badly for the ogre who, thanks to Little Thumb, slits the throats of his own seven daughters by mistake. Adding to the ogre's misery, Little Thumb not only manages to steal the ogre's magic boots but also cheats Mrs. Ogre out of all of their money. The tiny lad then returns home "where he was received with an abundance of joy" from his father who quickly realizes that he can probably retire from a career spent tying together bundles of twigs. One moral of this story is that you should not knife anyone in a darkened room where your kids are sleeping. Another appears to be that child-eating cannibals will not live happily ever after.

The brothers Grimm revisited a similar plot in *Hansel and Gretel*, which also detailed the abandonment of the young and the threat of cannibalism. The story begins with a concise and vivid portrayal of famine ("great scarcity fell on the land") but in the Grimms' tale, rather than an ogre's wife, a kindly old woman takes in the lost brother and sister. The hag, however, quickly reveals both her true witchy identity and her intentions after she locks Hansel in the stable. "When he is fat I will eat him," she cackles, and later, "Let Hansel be fat or lean, tomorrow I will kill him and cook him."

Other fairy tale writers also employed the cannibalism angle, most notably Englishman Benjamin Tabart (1767–1833) in his 1807 story *The History of Jack and the Beanstalk*. According to Maria Tatar, a leading authority on children's literature, Tabart, like Perrault and the brothers Grimm, based his tale on older tellings of the story. Although *Jack* existed in many versions, it is Tabart's that would become the model for subsequent adaptations, notably that of Joseph Jacobs, who compiled and edited five popular books of fairy tales in the late 19th and early 20th centuries.

In Tabart's story, Jack is "indolent, careless, and extravagant," and his actions bring his mother to "beggary and ruin." Trading in the family's milk cow to a stranger for a handful of seeds seems like a typical move for this lame incarnation of Jack, but of course things get interesting when his mother tosses the seeds away and an enormous beanstalk shoots up in hyper-bamboo fashion just outside their cottage. Climbing the ladderlike super stem, Jack meets a curiously tall woman and asks her for some breakfast. "It's breakfast you'll be if you don't move off from here," she tells him. "My man is an ogre and there's nothing he likes better than boiled boys on toast." But Jack is starving and, ignoring the danger, he convinces the wife to bring him back to her place for a bite. Soon enough, though, the ground is rumbling and Jack barely has time to jump into the oven before the Big Guy bursts in, reciting the most famous lines in all of ogredom:

> Fee-fi-fo fum,
> I smell the blood of an Englishman,
> Be he alive, or be he dead
> I'll have his bones to grind my bread.

Unimpressed, his wife tells him that he's probably dreaming, "Or perhaps you smell scraps of the little boy you liked so much for yesterday's dinner." Satisfied, the ogre has his breakfast before settling down for a nap. Jack, showing just how thankful he is to have been spared by the ogresse, promptly steals not only the couple's gold and a harp that plays itself but, because you can never have enough gold, he filches a goose that lays golden eggs. Next, after somehow hauling all of this loot down to the ground, Jack shows off his logging skills by cutting down the beanstalk just in time to send the ogre plummeting to his death.

In Joseph Jacobs's revised epilogue, a "good fairy" shows up and informs everyone that the giant had actually stolen the gold from Jack's late father. With the theft and killing justified, "Jack and his mother became very rich, and he married a great princess, and they lived happily ever after."[29]

In story after story, the Grimms, Perrault, and other fairy tale writers piled on scenes of cannibalism or, at the very least, its threat. In *Cannibalism and the Colonial World*, Marina Warner describes these collections as "the foundation stones of nursery literature in the West." As such, these stories served to reinforce the idea, for readers of all ages, that cannibalism was the stuff of nightmares and naughty children.

BEYOND THE HISTORIANS, playwrights, poets, and compilers of fairy tales, there were others who contributed to what became our culturally ingrained ideas about cannibalism. Three of the most influential were the writer Daniel Defoe, Scottish anthropologist Sir James George Frazer, and the Father of Psychoanalysis, Sigmund Freud.

Daniel Defoe (ca. 1660–1731) was a prolific author and perhaps the founding father of the English novel. Born in London as Daniel Foe, he eventually changed his name in an effort to construct an

29 Alfred Hitchcock used a similar technique on many occasions, appearing at the conclusion of his famous TV show to assure viewers (and censors) that the villain didn't really get away with his or her crime.

aristocratic origin from what had actually been a lower-class upbringing. It was a childhood during which young Daniel survived not only London's Great Plague in 1665 but also its Great Fire the following year. After abandoning an up-and-down career as a businessman, Defoe began writing books, pamphlets, and poems—many of them with a political bent. *Robinson Crusoe*, published in 1719, was his most famous work, and by the end of the 19th century it had become a worldwide phenomenon. Running through nearly 200 editions and translated into 110 languages, *Robinson Crusoe* has been abridged, pirated, spun off, and turned into an array of children's books, an opera, and several movies.

The plot of *Robinson Crusoe* follows the decades-long adventures of the shipwrecked title character as he struggles to survive on a tropical island, possibly based on the isle of Tobago. After establishing a relatively comfortable life for himself, Crusoe knows that the most serious threat to his safety comes from the man-eating savages who frequent the island. These wretches, the reader is informed, battled each other in canoes with the victors killing and eating their prisoners Carib-style. This grim predilection for murder and the consumption of human flesh is spelled out in sensational detail when the castaway comes upon the remains of a cannibal feast on the beach.

> I was perfectly confounded and amazed; nor is it possible for me to express the horror of my mind at seeing the shore spread with skulls, hands, feet, and other bones of human bodies; and particularly I observed a place where there had been a fire made, and a circle dug in the earth . . . where I supposed the savage wretches had sat down to their human feastings upon the bodies of their fellow-creatures.

After spewing his lunch (the suitable response of any civilized Englishman), Crusoe hurries back to his side of the island and his "castle" where, for the next two years, he fixates about "the wretched, inhuman custom of their devouring and eating one another up." Crusoe fantasizes gruesome plans for revenge, including one in which he sets off explosives under the cannibal cooking pit and another in which he blows off their heads from a sniper's nest. While brooding over his own obsession, Crusoe begins to doubt whether the savages actually knew that they were committing horrendous crimes. In a rare instance of 18th-century clarity regarding Columbus and those who followed him, Crusoe wonders whether killing the cannibals would "justify the conduct of the Spaniards in all their barbarities practiced in America, where they destroyed millions of these people."

Initially, the fictional castaway decides to steer clear of the savages, but he winds up killing one of them while rescuing "Friday"—

a cooking pot escapee, who is himself a cannibal. Once the main party of man-eaters departs, Crusoe and Friday return to the scene of the cannibal feast.

> The place was covered with human bones, the ground dyed with their blood, and great pieces of flesh left here and there, half eaten, mangled, and scorched. . . . All the tokens of the triumphant feast they had been making there, after a victory over their enemies.

After piling up the body parts and setting them ablaze, Crusoe observes that Friday "still had a hankering stomach after some of the flesh," and he lets the savage know in no uncertain terms that death awaits should he give in to his cravings. Friday quickly gets his own point across (presumably using a combination of miming and interpretive dance) that he "would never eat man's flesh anymore."

Years later, Crusoe and Friday come upon another cannibal banquet, and this time the next course appears to be Bearded White Guy. At this point, all of Crusoe's previously developed ideas about non-involvement in local customs are put to the test. But after downing a few shots of rum, the castaway and his lethal sidekick ("now a good Christian") wade in, and "Let fly . . . in the name of God," slaughtering 17 or 18 of the 21 man-eaters, with guns, swords, and a hatchet.

*Robinson Crusoe* had a major impact on readers all over the world. According to University of Sorbonne professor of literature Frank Lestringant, "Defoe's work is an effective contribution to the black legend of the Cannibals. It represents the normal English attitude towards them throughout the ages of discovery and colonization." In short, cannibalism was an abomination and cannibals were to be avoided, since God would ultimately sort out their fate. But if that didn't work, anyone who practiced man-eating could be enslaved or killed by any method, no matter how cruel or gruesome it might appear.

In 1890, Sir James Frazer (1854–1941) produced *The Golden Bough: A Study in Magic and Religion*, a massive, globe-spanning, comparative work on mythology and religion. Much of this material was accompanied by a hefty dose of archaeological support, and Frazer's enormously popular compendium of rites, practices, and religions greatly influenced the emerging discipline of anthropology. Throughout his magnum opus, Frazer discussed the practice of cannibalism and other barbarous customs. He also advised his readers not to be fooled into "judging the savage by the standard of European civilization."

Frazer pointed to several African tribes whose religious rites included "the custom of tearing in pieces the bodies of animals and of men and then devouring them raw. . . . Thus the flesh and blood of dead men are commonly eaten and drunk to inspire bravery, wisdom, or other qualities for which the men themselves were remarkable." According to Frazer, this type of cannibalism also took place among the mountain tribes of southeastern Africa, the Theddora and Ngarigo tribes of southeastern Australia, the Kamilaroi of New South Wales, the Dyaks of Sarawak, the Tolaalki of Central Celebes, the Italones and Efugao of the Philippine Islands, the Kai of German New Guinea, the Kimbunda of Western Africa, and the Zulus of Southern Africa.

During the first half of the 20th century, *The Golden Bough* influenced an array of major authors including Joseph Campbell, T. S. Eliot, Robert Graves, James Joyce, D. H. Lawrence, Ezra Pound, and William Butler Yeats. As previously mentioned, Frazer's work also became an enormously popular resource for the budding anthropologists who were beginning to trek into some of the most remote regions on the planet. Although each subsequent generation found flaws in Fraser's work or had to modify certain aspects of it, there is little doubt that his stance on the prevalence of cannibalism among indigenous people colored the mindset of many a fresh-faced anthropologist. As a result, when such groups were encountered, they were assumed to be savages whose behavioral repertoire would likely encompass all manner of strange rites including cannibalism. Contributing to this attitude was perhaps the most well known of these new anthropologists, Margaret Mead (1901–1978), who was famously quoted about some of the Pacific Islanders she was studying, "The natives are superficially agreeable but they go in for cannibalism, headhunting, infanticide, incest, avoidance and joking relationships, and biting lice in half with their teeth."

Anthropologists were not the only professionals talking about cannibalism and the primitive mind. For Sigmund Freud, the behavior denoted a pre-cultural stage of human development. In his appropriately named 1913 book, *Totem and Taboo*, Freud borrowed Darwin's concept of a patriarchal horde, where a single mature male ruled over a harem of females. Immature males ("the brothers"), who were forbidden to mate, also belonged to this primitive social group. Freud assumed that these fellows would be quite grumpy and, as such, he proposed that they were hot to initiate some revision of the prehistoric status quo. They did so by killing their father, thus putting an end to the patriarchal horde. "Cannibal savages as they were, it goes without saying that they devoured their victim as well as killing him"—each of the sons acquiring a measure of their father's strength. In order to commemorate the event, the brothers organized a totem feast, which Freud described as "mankind's earliest festival." This, though, was no ordinary party, since according to Freud, it marked the beginning of social organization, moral restrictions, and religion. Once cannibalism and its partner, incest, were abandoned, the group in question would be firmly on the road to civilization—a mindset that is highly reminiscent of the one espoused by explorers, missionaries, and early anthropologists as they encountered indigenous cultures. As Stony Brook University anthropologist Bill Arens wrote in 1979, "What could be more distinctive than creating a boundary between those who do and those who do not eat human flesh?"

Freud also went on to say that taboos (like cannibalism) represent forbidden actions for which there exist strong and unconscious predispositions—primitive urges buried deep within each of us. From a zoological perspective, these "primitive urges" can be seen as further evidence that we humans are (to paraphrase Stephen J. Gould) a part of nature, not apart from nature, and, as such, we still retain bits of an ancient genetic blueprint. We are also, however, of a lineage that has diverged greatly during our long evolution—and the more recently added or modified sections of our genetic code have seen us evolve us away from the behavior of spiders, mantises, and fish (though less so from our fellow mammals). Part of that divergence is that humans are cultural creatures, and for some of us the very underpinnings of our Western culture, starting with our literature, dictate that unless we are placed into extreme circumstances, certain practices, like cannibalism, are forbidden. But what about cultures in which those Western taboos were never established? Would they enact similar prohibitions on such behavior?

# 14: Eating People Is Good

*I think there is nothing barbarous and savage in that nation, from what I have been told, except that each man calls barbarism whatever is not his own practices; for indeed it seems that we have no other test of truth and reason than the example and pattern of opinions and customs of the country we live in.*

—Michel de Montaigne, *Of Cannibals*, 1580

*Mr. Chambers! Don't get on that ship! The rest of the book,* To Serve Man, *it's—it's a cookbook!*

—Patty, "To Serve Man" (*The Twilight Zone*), 1962[30]

One way to support a hypothesis that the origin, spread, and persistence of the Western cannibalism taboo can be traced along a line leading back to the Ancient Greeks, would be to find a culture with an extensive historical record that existed for millennia without the significant influences of Homer, Herodotus, and the Western writers who followed them.

Among many of the cultures that definitely weren't reading the

30 In one of the most well-known TV episodes of all time, humanity realizes too late that the intentions of an advanced race of alien visitors are somewhat less than benevolent.

Greek mythology (the Aztecs and Caribs come to mind), there is little if any proof as to their definitive stance on cannibalism. While there is a significant body of evidence regarding the Aztec practice of human sacrifice, which was clearly depicted in both carved inscriptions (glyphs) and bark paper books known as codices, there is no such consensus among historians that the Aztecs ever practiced cannibalism, especially on a large scale. And while a few Spaniards present in Mexico during the Aztec conquest provided written accounts of cannibalism, skeptics might question whether sources like "The Anonymous Conquistador" were reliable witnesses. Other tales of Aztec man-eating are similar to the secondhand reports of Carib cannibalism in that most of them were written by men who weren't present in Mexico until a decade after the Aztec empire had been destroyed—if they were present at all. Since there is no conclusive evidence the cannibalism was practiced by either the Aztecs or Caribs, we need to look elsewhere for a group not influenced by the Ancient Greeks.

Rather than focusing on one of the smaller linguistic groups, like the Wari' of Brazil or the soon-to-be-discussed Fore of New Guinea, to whom cannibalism was apparently not a taboo, I chose instead to examine a culture with a lengthy, exquisitely detailed, and well-studied history. That culture belongs to the Chinese, and while their enormous country may not have been completely isolated from Western influences, its leaders have been obsessive in maintaining what is apparently the world's longest unbroken historical record. How, then, did the Chinese deal with cannibalism—historically and in modern times? Are Western-style taboos present and, if not, what, if anything, does that tell us about humans as a species?

There is a general agreement among recent scholars that China has a long history of cannibalism.[31] The evidence comes from a range of Chinese classics and dynastic chronicles, as well as an impressive compendium of eyewitness accounts, the latter providing some unsparingly gruesome details about some of the most recent incidents.

In *Cannibalism in China* (1990), historian and Chinese cannibalism expert Key Ray Chong specified two forms of cannibalism: survival cannibalism, which might occur during a siege or famine; and learned cannibalism, which the author described as, "an institutionalized practice of consuming certain, but not all, parts of the human body." He went on to describe learned cannibalism as being "publicly and culturally sanctioned," making it synonymous with the term "cultural cannibalism."

As we have already seen, survival cannibalism was not unique among the Chinese, but the practice is worth discussing for several reasons—not the least of which was the frequency with which it occurred in China, coupled with a succession of governments whose responses varied from turning a blind eye to something close to official sanction. Perhaps the saddest and most surprising case (and the one with the greatest death toll) actually occurred in the mid-20th century, when starvation and cannibalism were only two aspects of a national calamity of unprecedented scope. It was a tragedy about which, until recently, much has been hidden from most Chinese citizens—and the world.

First, though, Chong's investigation provided three examples of siege-related cannibalism recorded in Chinese classical literature.

---

31 These authors include Key Ray Chong (1990), Jasper Becker (1996), Zheng Yi (1996), Lewis Petrinovich (2000), and Yang Jisheng (2008).

The oldest took place during a war between the states of Ch'u and Sung in 594 BCE and occurred in the Sung capital city. It was also notable because it was apparently the first time that starving Chinese began exchanging one another's children, so that they could be consumed by non-relatives—a practice made permissible by an imperial edict in 205 BCE. The other examples took place in 279 BCE in the besieged cities of Ch'u and Chi-mo, and in 259 BCE in the city of Chao. In the latter instance, soldiers defending a castle reportedly cannibalized servants and concubines, followed by children, women, and men "of low status."

In total, Chong's exhaustive research efforts yielded 153 and 177 incidents of war-related and natural disaster–related cannibalism, respectively. With no statistical difference in the numbers reported from the Han Dynasty (206 BCE–220 CE) to the Ch'ing Dynasty (1644–1912), incidents of cannibalism (in which varying numbers of people were consumed) seem to have been a fairly consistent occurrence throughout China's long history—until recently, that is. But rather than the decrease in reports of cannibalism one might expect to find in modern times, the opposite turns out to be true. The greatest number of cannibalism-related deaths in China came as a direct result of Mao Zedong's "The Great Leap Forward" (1958–1961), a disastrous attempt at utopian engineering.

This government program eventually morphed into what some consider the most far-reaching case of state-sponsored terrorism in the history of mankind. It also produced what may have been the worst famine in recorded history—a continent-spanning disaster in which at least 30 million, mostly rural, Chinese died of starvation. Those who wrote about the catastrophe often did so at their own peril, but what they uncovered was truly shocking. For example, in the 2008 book *Mubei* (*Tombstone*), Yang Jisheng wrote that famine-starved "people ate tree bark, weeds, bird droppings, and flesh that had been cut from dead bodies, sometimes of their own family members." The author, who lost his father to starvation, also believes that 36 million deaths is a more accurate number, although some estimates run as high as 46 million. In brief, this is how it all came about.

In an effort to transform China's primarily agrarian economy into a modern communist society based on industrialization and collectivization, Mao Zedong, Chairman of the People's Republic, ordered nearly a billion farmers to move from private farms to massive agricultural collectives. More often than not, these communal farms were run by government officials who had no farming experience at all. Making matters even worse, Mao had them institute an anti-scientific agricultural program that had sprung from the brain of semi-literate Soviet peasant Trofim Lysenko in the late 1920s. Lysenkoism (as it came to be called) initially led to a deadly purge of Russian scientists and intellectuals. Eventually it set the Soviet Union's agricultural system back at least 50 years and resulted in millions of starvation-related deaths.

Lysenko rejected an array of selective breeding techniques, especially those based on Mendelian genetics. Instead he proposed his own muddled version of Jean-Baptiste Lamarck's early-18th-century claim that environmental factors produced needs or desires within an organism that led to new adaptations. Lamarck's infamous giraffes, their necks stretching and lengthening in an effort to reach leaves in an ever-higher tree canopy, remain a common misconception of how variation in traits like color or size could be generated in any given population.

Lamarck, trained as a naturalist, believed that the giraffes willed these changes to occur—changes that would then be passed on to

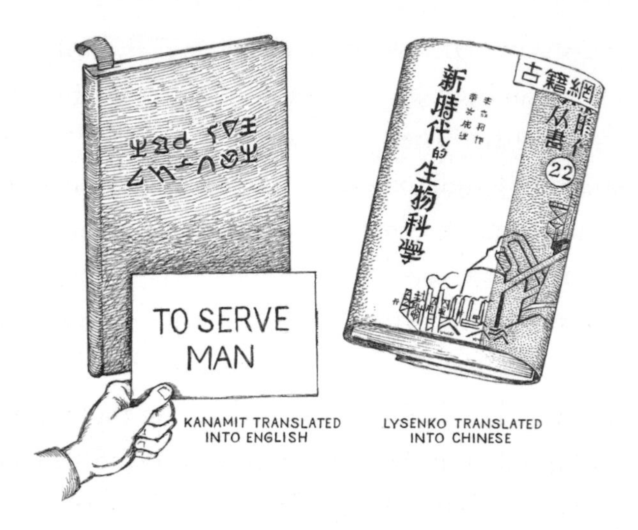


future generations. In Lysenko's interpretation, the organisms exhibiting these needs and desires were crop plants like corn, wheat, and vegetables. In that regard, Lysenko boasted that he could grow citrus trees in Siberia by cold-storing the seeds the previous year. These sorts of preposterous claims went on for decades, with those who questioned Lysenko's program either eliminated or afraid to make their voices heard. In the end, what Lysenkoism *did* prove was that reality does not yield to wishful thinking and truth cannot be established by a political party (or any other organization for that matter).

Not to be outdone by the Russians, Mao decided to install an "improved" version of Lysenko's agricultural program in China, and his so-called advances might have been comical if their consequences hadn't been so horrific. Instead of planting seedlings apart from each other, for example, Mao instructed that they be "close planted," since rather than competing for resources like water and nutrients, the tightly packed plants would, like the farmers Mao had packed into enormous communal farms, help each other to grow. The seedlings invariably died, although farmers were coerced into pretending that mature plants were so densely compacted that children could stand on them. Photographs depicting this "miracle" were achieved by having the kids stand on a bench, hidden from view.[32]

Combined with forced collectivization and a purge of anyone who seemed to have a clue about anything, the Great Leap Forward ended in catastrophe. Agricultural output (mostly grain) fell significantly, even though local officials grossly inflated their actual production numbers to curry favor with Mao. This imaginary surplus led to increases in government quotas, so that most of what was produced was immediately confiscated by the state and even exported. Meanwhile, the farmers starved, as did anyone else not living in a city. Farm animals were eaten, then pets, and finally the bodies of the dead, especially children. Foreign correspondent Jasper Becker (former Bureau Chief of the *South China Morning Post*), wrote: "Traveling around the region over thirty years later, every peasant that I met aged over 50 said he personally knew of a case of cannibalism in his production team. . . . Women would usually go out at night and cut flesh off the bodies, which lay under a thin layer of soil, and this would then be eaten in secrecy."

Critics of Mao's system were imprisoned or murdered, and thousands of farmers were accused of hoarding grain and were tortured to death in gruesome fashion. Fortunately, the Great Leap Forward, which was conceived as a five-year plan, was abandoned during year three. But although Chinese rulers looked the other way as starving populations consumed their dead, the cannibalism that took place during the Great Leap Forward was more of a necessity

---

32 Another of Mao's brainstorms led to war being declared on sparrows, with the subsequent success of farmers' efforts reflected by a concurrent increase in crop-munching insect populations.

than a choice. These instances of survival cannibalism (albeit on a massive scale) do not, therefore, answer the question of whether or not something like the Western cannibalism taboo also existed in China.

It is under the banner of learned cannibalism that the Chinese appear to have exhibited attitudes toward cannibalism that differed significantly from the Western taboos. For a start, author Key Ray Chong provided a list of circumstances that might lead to an act of learned cannibalism. These were "hate, love, loyalty, filial piety, desire for human flesh as a delicacy, punishment, war, belief in the medical benefits of cannibalism, profit, insanity, coercion, religion, and superstition." Some of these, Chong asserted, were uniquely Chinese.

As anyone who has ever visited China (or to a lesser extent, any big-city Chinatown) can attest, the Chinese consume a diverse range of creatures and their parts. Many of these, like scorpions and chicken testicles, fall outside the range of typical Western diets and, as writer Maggie Kilgore pointed out in 1998, some, like rats, snakes, shellfish, and things with paws, are specifically banned by Judeo-Christian law. Without our long list of forbidden foods, it's not a surprise that the Chinese felt less strongly about consuming other humans.

Throughout their long history, body parts were such important ingredients in Chinese cuisine that Key Ray Chong devoted a 13-page chapter to "Methods of Cooking Human Flesh" with a subheading entitled "Baking, Roasting, Broiling, Smoke-drying, and Sun-drying." And rather than an emergency ration consumed as a last resort, there are many reports of exotic human-based dishes prepared for royalty and upper-class citizens. T'ao Tsung-yi, a writer during the Yuan Dynasty (1271–1368), wrote that "children's meat was the best food of all in taste" followed by women and then men. In *Shui Hu Chuan* (*The Tales of Water Margins*), a novel written in the 12th century, there are numerous references to steamed dumplings stuffed with minced human flesh, as well as a rather nonchalant regard by merchants and customers over the sale of human meat.

Even if epicurean cannibalism isn't limited to the Chinese, the extent to which it was set down in detail certainly was. Amidst information on "five regional cuisines" (Szechwan, Canton, Fukien, Shantung, and Honon), the *San Kuo Yen Ki* (*Dramatic Epic of the Three Kingdoms*), written in 1494, contained "many examples of steaming or boiling human meat." Prisoners of war were preferred ingredients, but when they ran out (figuratively or literally), General Chu Ts'an's soldiers seized women and children off the street, killed them, and then ate them. As recently as the 19th century, executioners reportedly ate the hearts and brains of the prisoners they executed, selling whatever cuts were left to the public.

Widespread epicurean cannibalism was still taking place in the late 1960s during the Cultural Revolution, although there was certainly an element of terror involved. Chinese dissident journalist Zheng Yi wrote the following in 2001:

> Once victims had been subjected to criticism, they were cut open alive, and all their body parts—heart, liver, gallbladder, kidneys, elbows, feet, tendons, intestines—were boiled, barbecued, or stir-fried into a gourmet cuisine. On campuses, in hospitals, in the canteens of various governmental units at the brigade, township, district, and country levels, the smoke from cooking pots could be seen in the air. Feasts of human flesh, at which people celebrated by drinking and gambling, were a common sight.

Another form of cannibalism in China had nothing to do with persecution and punishment. Key Ray Chong reported that, "children would cut off parts of their body and make them into soup to please family members, particularly their parents." This last example, and many others like it, led him to study what he considered a truly unique aspect of learned cannibalism among the Chinese—its association with the Confucian philosophy of filial piety. In general terms, filial piety is a highly regarded virtue in which it is the duty of younger family members to demonstrate respect, obedience, and care for one's parents and elderly family members. In this case, however, it refers to an extreme act of cannibalism-related self-sacrifice, with relatives providing parts of their own bodies for the consumption and benefit of their elders.

Although by no means meant to be a complete list, Chong came up with a total of 766 documented cases of cannibalism-related filial piety, spanning a period of over 2,000 years. The practice took place primarily between sons and fathers, sons and mothers, and daughters and mothers.[33] The most commonly consumed body part was the thigh, followed by the upper arm, both of which were prepared in a rice porridge called congee. Far less frequent, but recorded nonetheless, were instances in which a young person volunteered a part of their liver, breast, finger, or even eyeball.[34]

In each case, the practice was intended to provide nutrition to a starving loved one or as a treatment of last resort, to afford the sufferer some medical benefit. The concept of medicinal cannibalism will be discussed further in Chapter 15.

---

33 Rarely, this exchange took place between daughters-in-law and fathers-in-law, and between daughters-in-law and mothers-in-law.

34 Although "an official edict in 1261 banned cutting out the liver or plucking out the eyeballs."

So is there any link between the practice of filial cannibalism in humans and that exhibited in the animal kingdom by species like mouth-brooding cichlids? One similarity is that, in both instances, the parent gains a benefit at the expense of the offspring. In humans, though, culture dictates that the offspring consciously initiate the act of filial cannibalism. Alternately, in animals it's the parents that do the initiating—regarding their offspring as a handy food source should their numbers drop too low to expend further energy on them, or when other forms of nutrition are unavailable.

In addition to the historical record of cannibalism contained within China's dynastic histories, the behavior in its various incarnations is also abundantly documented in plays, poems, and other works of fiction. For example, the 15th-century play, *Shuang-zhong ji* (*Loyalty Redoubled*) tells of a general coming up with the idea of turning his concubine into soup to feed his besieged and starving troops. Happily, for the general at least, the concubine volunteers for this duty, thus sparing the general from having to murder an innocent woman. The concubine's devotion spurs the soldiers to fight on, which leads another servant (this one a boy) to volunteer his own body.

According to numerous sources, then, the practice of cannibalism in China was more or less accepted as a necessity during times of famine, as a right to be exercised during warfare and acts of vengeance, and as a way of honoring one's relatives. And even though cannibalism wasn't something the majority of Chinese ever looked forward to, the behavior apparently never had the same stigma attached to it that it did in Western cultures.

Many cultural and physical anthropologists vehemently disagree with writers such as Bill Arens that examples of cultural or ritual cannibalism were made up. They cite reams of ethno-historical data as well as physical evidence as proof that this type of behavior occurred across the entire span of human existence. But whether various acts of culturally sanctioned cannibalism existed or not (and it seems absurd to consider that they never did), the fact remains that for the vast majority of Westerners, our feelings regarding the practice have resulted (at least in part) from our exposure to a long list of influential writers beginning with the Ancient Greeks and extending into the 21st century.

In China there were no such widespread taboos regarding the behavior—which was carried out for a variety of reasons. Eventually, though, Western civilization came to dominate much of what the world saw and often emulated and, as Western influences made their way into Chinese society, the Greek and Judeo-Christian abhorrence for cannibalism began to rub off—at least outwardly. This may explain the current silence and stigma in China about the survival-related cannibalism practiced during China's Great Leap Forward.

But if I've given you the impression that cannibalism did not occur in the West, that would be an error. It was actually a common practice in places like Europe, where it was carried on in various forms into the 20th century. It is also being practiced today here in the United States.

# 15: Chia Skulls and Mummy Powder

*The ancients were very eager to embalm the bodies of their dead, but not with the intention that they should serve as food and drink for the living as is the case at the present time.*

—Ambroise Paré (1582), cited in *Consuming Grief* by Beth Conklin

As secretaries and colleagues began to move into Dr. Bill Arens's office with greater frequency, I could tell that my interview with the anthropologist was drawing to a close. I decided to go for broke, pressing him for some acknowledgement that ritualized cannibalism existed . . . somewhere.

"So Dr. Arens, which example of ritual cannibalism do you find hardest to refute and why?"

"Well," he replied, "it depends on the definition of 'ritual cannibalism.' Because if you think that grinding up body parts and using them for medicinal purposes is ritual cannibalism, then I would find that the most difficult to reject."

I should mention that in 1998 Arens had previously acknowledged that the existence of rituals involving "the ingestion of culturally processed human body parts is open to further consideration." Referring to a reported instance of bone ash cannibalism by the Amahuaca of the Peruvian-Brazilian border, Arens admitted that he had been "unreasonable" in denying its occurrence.

"But you don't consider that cannibalism, do you?"

"I do. I do consider that cannibalism," Arens said (and I'm sure he was taking some satisfaction at my shocked expression). "I think if people in South America grind up bones and ingest them and people in America grind up organs and ingest them, then it's cannibalism. But I think you either have to say that this is not cannibalism or they're both cannibals. You can't choose between them. And so if you accept that [this type of behavior] is cannibalism, then *that* would be the most difficult to reject because it takes place—it has taken place. Some things you call 'ritual cannibalism' are impossible to deny or reject."

I left the interview with a new respect for Arens, who'd taken major heat from his colleagues for claiming that there was no first-hand evidence that ritualized cannibalism ever had existed in any society. But rather than trumpeting a belief that ritual cannibalism never occurred, what Arens had actually done was to knock down a hornet's nest and then give it a kick, in order to make some extremely relevant points.

One of these concerned the racism inherent in applying the "cannibal" tag to a cultural group—a practice undertaken by a long list of flag-planting invaders and those who accompanied and followed them. Keenly aware that the Western cannibal taboo had reduced these indigenous inhabitants to subhuman status, the invaders were able to justify the use of any form of behavior (no matter how inhumane) in order to subjugate and, in many instances, utterly destroy cultural groups of every size, wherever they were encountered. Arens and others like him have forced us to revise our views on the Age of Discovery as well as the explorers we have, perhaps, too long regarded as heroes. One needs only to look at the growing push to celebrate Indigenous Peoples Day instead of Columbus Day to see this revision in action.

Arens was also at least partially responsible for an increase in the scientific rigor with which fellow researchers explored cannibalism-related topics: "although the theoretical possibility of customary [i.e., ritual] cannibalism cannot be dismissed," he wrote in *The Man-Eating Myth*, "the available evidence does not permit the facile assumption that the act was or ever has been a prevalent cultural feature." It's probably no coincidence that since the publication of his book in 1979, anthropologists have developed and adhere to a clear-cut set of criteria when attempting to validate claims of cannibalism in their study groups, whether those groups are extinct or extant. The overall effect has been that the majority of anthropologists now believe that ritual cannibalism was practiced by fewer cultural groups throughout history than was previously thought. In all likelihood, this has resulted from an increased degree of scrutiny being applied to any proposed evidence of cannibal-related behavior.

I am convinced that these two outcomes were Arens's true goals. As for those who might wonder why he had ruined his reputation—in reality, he hadn't. He has plenty of 21st-century supporters who now agree with the "read-between-the-lines" contributions he made in 1979. And even his detractors can't help plastering his name all over their own papers. I'm certain Dr. Arens is amused.

Arens's examples of consuming pulverized human bones or organs in order to treat some malady fall under the general heading of medicinal cannibalism, which is, once you consider it, a form of ritual cannibalism. But however it's classified, the practice is as interesting as it is little known. It turns out that medicinal cannibalism was once widespread throughout Western culture, although reference to it has essentially disappeared from the historical record. The same, however, cannot be said for the Chinese, whose literature, medical texts, and historical accounts span over 2,000 years and contain detailed descriptions of the preparation and use of body parts as curatives.

Scholar Key Ray Chong wrote that the first documented use of organs and human flesh to cure diseases in China took place during the Later Han period (25–220 CE) and that medicinal cannibalism became increasingly popular beginning in the Tang Dynasty (618–907 CE), when it became associated with filial piety. By the end of the Ch'ing Dynasty (1644–1912), Western missionary doctors were reporting that the Chinese medical treatments included the consumption of "the gall bladder, bones, hair, toes and fingernails, heart and liver." Thomas Chen, a pathology professor at the New Jersey Medical School, tells us that "nail, hair, skin, milk, urine, urine sediments, gall, placenta and even flesh" were used in China for a variety of medicinal purposes.

But what about the reports of medicinal cannibalism in Europe, some of it taking place into the 20th century? Considering how outraged the Spanish were upon learning about the man-eating behavior of the indigenous people of the Caribbean, one might assume that cannibalism of any kind would have been frowned upon. But that was certainly not the case. As it turns out, many Renaissance-enlightened Christians from Spain, England, France, Germany, and elsewhere turned to medicinal cannibalism to treat a long list of problems. From kings to commoners, Europeans routinely consumed human blood, bones, skin, guts, and body parts. They did it without guilt, though it often entailed a healthy dose of gore. They did it for hundreds of years. Then they made believe that it never happened.

Perhaps the most commonly consumed human tissue is blood—a substance that has, until fairly recently, been misunderstood. Until the 20th century, most of what we knew (or thought we knew) about blood could be traced to the 2nd century Greek physician Claudius Galenus, known as Galen. Physician to the Roman gladiators, Galen stressed the importance of four bodily humors: blood, black bile, yellow bile, and the smoker's favorite, phlegm. According to Galen, keeping the body's humors in balance was the key to good health, both mental and physical. Unfortunately, this doctrine would become the party line for medical practitioners for well over a thousand years, with Galen's followers routinely involved in serious bouts of bleeding, gorging, and purging (the latter from both ends).

Since Galen believed that blood was the most important of the humors, bloodletting, usually initiated with a blade called a lancet, was prescribed to treat everything from fever and headaches to menstruation. Some of this blood, though, ended up back in the patient, where it was consumed to treat epilepsy. So popular was this practice that public executions routinely found epileptics standing close by, cup in hand, ready to quaff their share of the red stuff.

But drinking down blood while it was hot and fresh was not the only way to take one's medicine. It was also dried and made into powder or mixed into an elixir with other ingredients. Most interesting to me was that consuming blood turned out to be far more than a medieval folk remedy, as evidenced by the fact that English physicians were still prescribing it as late as the mid-18th century.

Although Galen's mistaken views would dominate the field of medicine for 1,500 years, the continued popularity of medicinal cannibalism can be primarily attributed to the rise of an alternative medical doctrine initiated by Philippus Aureolus Theophrastus Bombastus von Hohenheim. Better known as Paracelsus (1493–1541), the Swiss physician is considered by some to be the Father of Chemical Pharmacology and Therapeutics, due to his pioneering use of substances like mercury, sulfur, and opium. He has also been called the world's first toxicologist. Still, many of Paracelsus's beliefs were founded on bizarre magic like alchemy, often infused with astrological mumbo-jumbo. Long after his death, his followers touted a medical philosophy that stressed the healing powers of the human body, but not in the manner we're familiar with. Rather, Paracelsian physicians often prescribed medications made from human body parts. For example, they might give epileptics a potion containing powdered human skull, a substance that they believed did double duty as a cure for dysentery.

Richard Sugg, author of the 2011 book *Mummies, Cannibals and Vampires: The History of Corpse Medicine from the Renaissance to the Victorian*, writes that every imaginable body part was used, including "human liver . . . oil distilled from human brains, pulverized heart, bladder stones, warm blood, breast milk, and extract of gall." Also popular in medicinal concoctions were bones, flesh, and fat, the latter applied to wounds or taken internally to treat rheumatism.

During the European Renaissance, the popularity of medicinal cannibalism may have begun within the great unwashed masses, but it was soon adopted as de rigueur by the enlightened, pious, and well-heeled. Upper-class types and even members of the British Royalty "applied, drank or wore" concoctions prepared from human body parts, and they continued to do so until the end of the 18th century. According to Sugg, "One thing we are rarely taught at school yet is evidenced in literary and historic texts of the time is this: James I refused corpse medicine; Charles II made his own corpse medicine; and Charles I was made into corpse medicine."

Additional high-profile advocates of medicinal cannibalism included Francis I (King of France), Jacopo Berengario da Carpi (Italian anatomist), John Donne (poet and priest), Francis Bacon (pioneer of the scientific method), John Banister (surgeon to Elizabeth I), John Hall (physician and Shakespeare's son-in-law), and Robert Boyle (natural philosopher, chemist, and inventor).

With an ever-increasing demand for human body parts, the popularity of public executions rose dramatically in the 17th century. The already-gruesome events became even gorier as the choicest cuts were harvested from prisoners, often while they were still alive. In what was described as a typical account from 1660 London, a prisoner has his "privy members cut off before his eyes" and "his bowels burned." He was then decapitated and his head set on a pole. Finally his body was cut into quarters, perhaps to maximize its value when divvied up for medicinal purposes or displayed "upon some of the city gates."

Human skulls not ground into powder were often left out in the air, where they served as the substrate for "skull moss"—a curative applied topically to stem bleeding and to treat disorders of

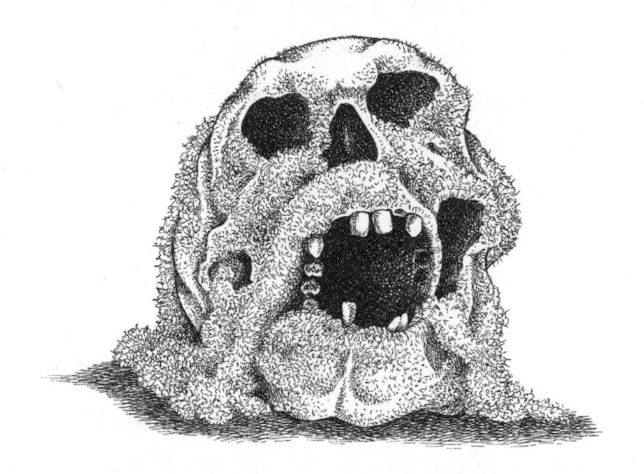

the head. Researcher Paolo Modenesi believes that the term actually refers to a taxonomic assemblage of mosses and lichens.[35] Renowned for their ability to thrive on bare rock, these organisms had little difficulty growing on the calcium-rich crania. Ideally, the moss from the skulls of hanged men was preferred but, according to naturalist and philosopher Robert James (1703–1776), Paracelsus believed that moss grown on the bodies of the unburied dead was quite acceptable. One set of directions showed just how easy it was to transform a skull into the medieval equivalent of a Chia Pet. The recipe called for the moss collected from a meadow in April to be dried and ground into a powder. This was sprinkled with a strong, sweet wine to form a paste, which was spread over "the cranium of a carcass that had been broken on the wheel." Gardeners were

---

35 Lichens are the results of a symbiotic relationship between a fungus and either an algae or a cyanobacterium. The fungi reap the energy benefits of photosynthesis while providing protection to the photosynthetic algae or bacterium.

advised to place their Chia skulls in the sun and warned to take them indoors when it rained.

The lichen *Usnea humana*, was also the main ingredient in a preparation called *unguentum armarium*, or "weapon ointment." This preparation, which also contained human blood and fat, was employed in a bizarre medical treatment known as hoplochrisma (*oplon* = weapon, *chrisma* = salve). Those administering this procedure might bandage a wound, but would otherwise leave it untreated. They would use the ointment itself on either the weapon that had caused the injury (if available) or a wooden facsimile of it. Given the fact that hoplochrisma had no side effects, it might be classified as one of the most effective treatments available at the time, even if the benefits were simply a result of the placebo effect.

Perhaps the most famous example of European medicinal cannibalism was the curious custom of pulverizing mummies to produce a substance known as mumia. This was either consumed (often as a drink ingredient) or applied topically as a salve or in a cloth compress. Mumia was used in the treatment of ailments ranging from epilepsy and bruising to hemorrhaging and upset stomachs. The problem was that there were only a limited number of Egyptian mummies being sent to Europe, leading to shortages and legions of grumpy mummy fans. In response, a thriving cottage industry popped up to supply ersatz mumia. Reportedly, by the end of the 17th century the quality of bootleg mummy had gotten so bad that buyers were advised to "choose what is of a shining black, not full of bones and dirt, and of a good smell."

There were, however, some high quality "artificial mummies" to be had (or at least a recipe for their production), as anthropologist

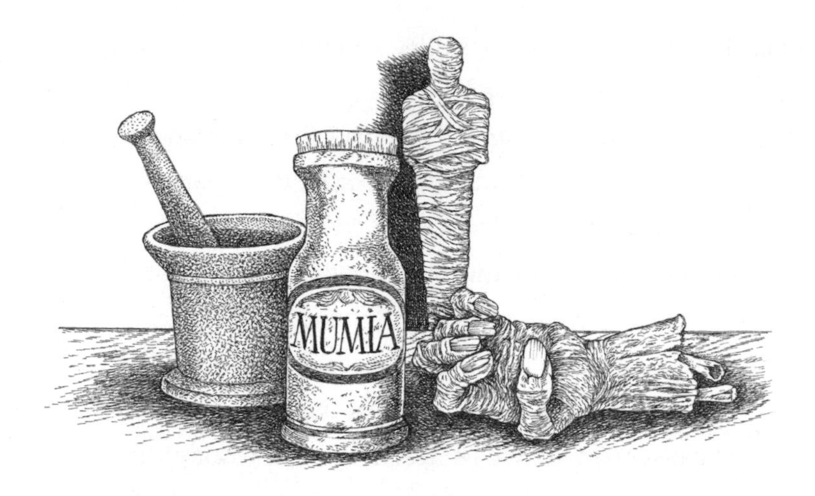


Karen Gordon-Grube uncovered in the official *London Pharmacopoeias* of the 17th century:

> [The Paracelist Oswald] Croll recommended that mummy be made of the cadaver of a redheaded man, age 24, who had been hanged. The corpse was to lie in cold water in the air for 24 hours, after which the flesh was cut in pieces and sprinkled with a powder of myrrh and aloes. This was soaked in spirit of wine and turpentine for 24 hours, hung up for 12 hours, again soaked in the spirit mixture for 24 hours, and finally hung up to dry.

In an interesting turn of fate, the popularity of grinding up mummies for medicinal purposes may have started because of a mistranslation. Apparently Arabs often used the petroleum-based substance we call tar or bitumen as an adhesive and to staunch wounds. Their word for this material was *mumia* but it also became their word for the mummified human remains they discovered after taking over Egypt in the 6th century CE. The Arabs mistakenly believed the mummies to have been prepared with bitumen during the preservation process. Centuries later, Europeans heard about the medical benefits of *mumia*. Unfortunately, they wound up hoarding *mumia*—the dried-up dead guys, rather than *mumia*—the tarry stuff. Either the locals never figured out the screw-up (which seems highly unlikely) or they simply never bothered to tell the Europeans about it. As a consequence, mummy powder was available at the Merck Pharmacy in Darmstadt, Germany, until 1908. Listed as *mumia vera aegyptica*, it sold for 17.50 marks/kg.

Essentially, then, as European adventurers, missionaries, and colonists were condemning the indigenous people they encountered for practicing cannibalism, their own rulers and countrymen in Europe were consuming human body parts to a degree and at a rate that would have made Hannibal Lecter proud. Until, that is, they stopped.

Richard Sugg, the foremost expert on the topic, believes that the practice of medicinal cannibalism was abandoned because of "the rise of Enlightenment attitudes to science, superstition, and the general backwardness of the past; a desire to create a newly respectable medical profession; a changing attitude towards hygiene, the body and disgust; and the radically changed nature of the human body itself." The latter development Sugg described as "a more mechanized model of the human body: an entity now drained (at least for the educated) of its animistic, essentially cosmic vitality." In short, its spirit and soul were gone. Although not in every case.

In 2002, stories began circulating that Keith Richards had mixed his dad's ashes with some cocaine and snorted them shortly after Bert Richards's death that year. Not so, replied "Keef," "after having Dad's ashes in a black box for six years, because I really couldn't bring myself to scatter him to the winds, I finally planted a sturdy English oak to spread him around. And as I took the lid off the box, a fine spray of his ashes blew out onto the table. I couldn't just brush him off, so I wiped my finger over it and snorted the residue. Ashes to ashes, father to son."

In a more widespread, though no less personal example, another form of medicinal cannibalism was experiencing an American revival in the 21st century.

# 16: Placenta Helper

*It gave me the wildest rush.*

—"The Placenta Cookbook," *New York* magazine, 2011

Thumbing through an issue of *New York* magazine several years ago, I stopped at what appeared to be a recipe-related article by the alliteratively named Atossa Araxia Abrahamian. Across a two-page spread was a photo of what looked to be an über-veiny roast beef, bobbing in a black enameled stew pot. Floating alongside the softball-sized blob of meat was a sliced jalapeño, a walnut-sized chunk of ginger, and a halved lemon. I read the title of the article, "The Placenta Cookbook," realizing that the main ingredient in this particular dish wasn't beef at all.

Of course I kept on reading.

Throughout the article there were quotes from several women, each of whom was enthusiastic about having consumed her own placenta. "Perfect," "beautiful," "precious," gushed one woman who had tried it. I also learned that while some moms preferred their placenta raw, others favored placenta-flavored smoothies, placenta jerky, and even a particularly apt version of a Bloody Mary. For those turned off by the idea of turning their placentas into a food item, or even handling the organ themselves, there were professional placenta-preparers who could be hired to procure the placenta from the hospital or accept its delivery by mail. These folks would then transform it into a bottle of encapsulated nutritional supplements, thus placing the whole placentophagy experience on a par with popping a Flintstones vitamin (that is, of course, assuming the character-shaped pills were actually made out of Pebbles or Bamm-Bamm). On that note, the article included a handy section (more color photography) for those readers wondering how these "happy pills" were made.

STEP 1: DRAIN BLOOD AND BLOT DRY . . .

STEP 5: GRIND IN BLENDER AND POUR PLACENTA POWDER INTO PILL CAPSULES.

From a biological viewpoint, the first question is, obviously, what is the function of a placenta? As a zoologist I was interested in determining what other mammals (besides white middle-class Americans) ate their own placentas and why they did it.[36] There were claims from some midwives and alternative-healthcare advocates extolling the therapeutic benefits of placenta consumption. What were these supposed benefits and, more importantly, was there any proof that they existed? I was also interested in determining whether additional human body parts had been (or were being) ingested for medicinal reasons. Finally, there was the question that had been worming around my brain from the very moment I'd finished the Abrahamian article: What did placenta taste like? Given its bloody, glandular appearance, my initial guess was calves' liver.

As it turned out, I was wrong.

---

36 In a 2013 study conducted by researchers at UNLV, 198 women who had "ingested their placentas after the birth of at least one child" were surveyed: 93 percent were white, 91 percent were from the U.S., 90 percent were married, and 58 percent reported a household income of more than $50,000 per year.

But first things first.

Advocates of placentophagy are likely to find it more than coincidental that the word placenta is derived from the Greek *plakous*, or "flat cake." The Latin term *placenta uterina*, or uterine cake, was coined by the Italian anatomist, Realdo Colombo. Tempering any culinary-related enthusiasm is the likelihood that the 16th-century scientist was referring to the flattened or slab-like nature of the roughly discus-shaped organ and not its potential as a base for chocolate frosting and candles.

The placenta is the organ that gives more than 9 out of 10 mammals (or roughly 4,000 species) their name—placental mammals. Also known as eutherians, the oldest placental mammals date from around 160 million years ago. Mouselike, they generally kept out of sight while the dinosaurs ran the show. But using their relatively larger brains and enhanced thermoregulatory abilities, they carved out slender niches of their own. Then, approximately 65 million years ago, as the planet underwent cataclysmic environmental changes (including those initiated by a six-mile-wide meteor striking near the current Yucatan Peninsula), the mammals hunkered down and survived. Once the dust settled, and many of our favorite kiddie toys had taken on new roles as fossils, the mammals exploded in diversity, speciating and spreading rapidly across a planet suddenly filled with evolutionary opportunity—a.k.a. open jobs.

Within approximately 10 million years of the dinosaurian demise, mammals diversified into all of the existing mammalian orders—rodents, bats, carnivores, primates, etc. Some took to the air while others returned to the water—each group evolving and passing on its own suite of adaptations, like wings or fins, to supplement basal mammalian characteristics like hair and bigger brains. Many of these species went extinct themselves. Others thrived, eventually outcompeting many of the non-mammalian vertebrates that had also survived the great die-off. And except in isolated regions like Australia and South America (which were effectively isolated from the expansion of the terrestrial placentals), the eutherians even outcompeted the older, non-placental mammals—the marsupials and the egg-laying monotremes.[37]

The organ that gives placental mammals their name is transient in nature, undergoing its entire rapid development only after conception. The tissue is derived from the fetus, as opposed to the mother, and in humans it has an average diameter of about nine inches. Thickest at its center (up to an inch), it thins out toward the edges and weighs in at just over a pound. The placenta functions as an interface between the mother and the developing fetus, connecting it to the mother's uterine wall but acting as a buffer as well. The organ itself is richly vascularized, which gives it its dark reddish-blue to crimson color, which relates to the placenta's life-support function: carrying oxygen and nutrients from the mother to the placenta and then from the placenta to the fetus via the umbilical artery. Structurally, most of the placenta is composed of cells called trophoblasts, which have a dual role. Some form small cavities that

---

37 Currently, there are 5 species of monotremes (4 echidnas and the platypus) and 334 species of marsupials. The latter are commonly referred to as "pouched mammals," although a pouch, or marsupium, is not a requirement for entry to the marsupial club. What all marsupials *do* share is a short gestation period, after which the fetuslike newborn takes a precarious trip from the vaginal opening to a teat (usually found within the marsupium). Upon finding one, the tiny creature latches on for dear life, and continues what is essentially the remainder of its fetal development for additional weeks or even months.

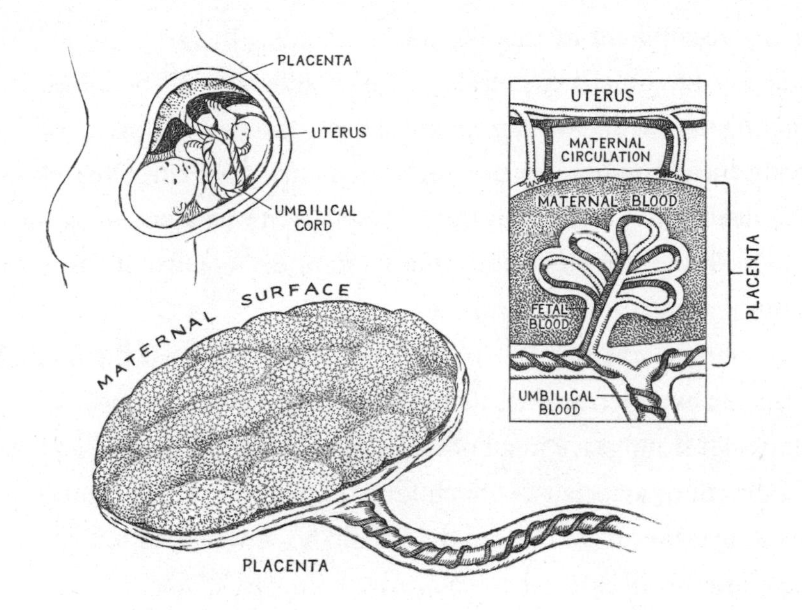


fill with maternal blood, thus facilitating the exchange of nutrients, waste, and gases between the fetal and maternal systems. Other trophoblasts specialize in hormone production. Waste products and carbon dioxide travel from the fetus back to the placenta via the umbilical vein. A sheath of connective tissue binds and protects both umbilical vessels, and together the entire structure is known as the umbilical cord.[38]

The placenta has additional functions, which include the production and release of several hormones, including estrogen (which maintains the uterine lining during pregnancy) and progesterone (which stimulates uterine growth as well as the growth

38 The word umbilical is Latin for "navel" or "middle." Blood from the umbilical cord is rich in stem cells and so it is sometimes collected and "banked," to potentially be used down the road to treat a number of blood-related disorders, including leukemia and lymphoma.

and development of the mammary glands). It also prevents the transfer of some, but not all, harmful substances—blood-borne pathogens for one—from the mother to the developing fetus. Finally, the placenta secretes several substances that effectively cloak the developing fetus from the mother's immune system—similar to the way in which immunosuppressant drugs prevent the body from rejecting a transplant.

Given its essential role in fetal development, what the human placenta experiences after delivery must surely be the most precipitous fall from grace in all of Organdom. Expelled by the uterine contractions associated with childbirth, this complex and amazing structure goes from revered mammalian namesake to biohazardous "afterbirth" faster than you can scream "PUSH!"

In the vast majority of mammals, though, the newly delivered placenta serves one last purpose.

IN 1930, PRIMATOLOGIST Otto Tinklepaugh took a break from his groundbreaking study on chimpanzee vaginal plugs to coauthor an article on the birth process in captive rhesus macaques (*Macaca mulatta*). He noted that the monkeys, and just about every other terrestrial mammal except humans and camelids (camels, llamas, alpacas, and their relatives) consumed their own placentas after giving birth. More recently, the behavior in the animal kingdom has been studied in rodents, lagomorphs (rabbits and their kin), carnivores, primates, and most artiodactyls (hoofed mammals).

Mark Kristal is the world's foremost authority on placentophagology, and until recently, he may have been the *only* expert on the topic. A SUNY Buffalo professor emeritus of psychology, Kristal's research began more than four decades ago. His work supports the hypothesis that, since placenta-eating has been observed in such a variety of mammals, it probably evolved independently and in response to one or several survival-related problems.

Researchers initially posited that eating the placenta kept the birthing area sanitary while eliminating smells that might attract predators. The fact that chimps giving birth in the trees hung around to eat their placentas instead of simply moving off (or flinging them down on some cheetahs), suggested that a new hypothesis was needed. Answering the call, dietary researchers suggested that placentophagy replenished nutritional losses associated with late-stage pregnancy and delivery. Endocrinologists hypothesized that moms might be acquiring (and replenishing) hormones present in the afterbirth. Other researchers suggested that placentophagy sated a mother's hunger after the delivery, or that placentophagy demonstrated the new mothers' tendency to develop "voracious carnivorousness" after giving birth.[39]

Kristal wasn't buying it, though. His skepticism for any of those proposed functions stemmed from the complete absence of any valid research on the topic. He and his colleagues set out to investigate placentophagy in non-humans experimentally—in this case, lab rats. As happens frequently, the results of their experiments supported none of the earlier hypotheses. Kristal did suggest, though, that the previously proposed functions of placentophagy might provide secondary benefits, if they existed at all.

"The main thing that we found during our studies turned out to be an opiate-enhancing property," Kristal said. He explained that

---

39 I discounted unpublished reports that the male author of the voracious carnivore hypothesis was found choking on a disposable diaper.

placenta consumption by new rat moms appeared to increase the effectiveness of natural pain-relieving substances (opioid peptides) produced by the body. He added that that these enhanced analgesic effects lasted throughout the birth interval between individual "kittens" in a litter—an important point since rats generally give birth to seven to ten individuals.[40]

Kristal also told me that the results of a second set of experiments linked afterbirth consumption (by rat moms) to a form of reward for parental care. Briefly, the central nervous system, pituitary gland, digestive tract, and other organs secrete pain-blocking peptides like endorphins, enkephalins, and dynorphins, which have been used to explain terms like "runner's high" and "second wind," as well as the phenomenon in which gravely wounded individuals report feeling little or no pain. Kristal's experiments indicated that those mothers who consumed their afterbirth received enhanced benefits from these natural painkillers, essentially getting an anesthetic reward for initiating maternal behavior like cleaning their pups.

I asked Kristal how long humans had been practicing placentophagy and how widespread the practice was. "I haven't discovered any human cultures where it's done regularly," he told me. "When placenta-eating is mentioned, it's usually in the form of a taboo. You have cultures saying things like 'Animals do it and we're not animals, so we shouldn't do it.'"

In 2010, researchers at the University of Nevada, Las Vegas, searched an ethnographic database of 179 preindustrial societies for any evidence of placenta consumption. Searching for the terms

---

40 Yes, rat babies are known as kittens (which should make dog lovers smile). The largest kitty litter I was able to uncover is 26—presumably a tough number for the 14 baby rats that couldn't immediately latch on to a nipple.

"placenta" and "afterbirth" in the electronic Human Relations Area Files (eHRAF), described as "the gold standard for cross-cultural comparative research," they found 109 references related to the special treatment and/or disposal of placentas. The most common practice (seen in 15 percent of accounts) was disposal without burial (examples include "throwing it into a lake"), followed by "burial" (9 percent). The latter narrowly beat out my personal favorite "hanging or placing the placenta in a tree" (8 percent). What the UNLV researchers did not find was a single instance of a cultural tradition associated with the consumption of placentas by moms—or anyone else, for that matter.

Considering the ubiquitous nature of placentophagy in mammals, including chimps, our closest non-human relatives, I was surprised they were unable to find at least one culture somewhere where placentas were regularly eaten. I mentioned to Kristal that I'd run across an example of placentophagy in the *Great Pharmacopoiea of 1596* (a go-to guide for many New Yorkers seeking medical advice), wherein Li Shih-chen recommended that those suffering from *ch'i* exhaustion (whose embarrassing symptoms included "coldness of the sexual organs with involuntary ejaculation of semen") partake in a mixture of human milk and placental tissue.

"It is an ingredient in herbal medicine," Kristal said. "In fact, there are a lot of placentophagia/midwife/doula websites where two things come up repeatedly. One—the benefits that I found in my research, which we never extrapolated to humans, and two—the [mistaken] idea that it's been done for centuries in China."[41]

---

41 A doula (from the Ancient Greek for "female servant") is a non-medical person who assists the mother before, during, and after childbirth. After

"So has it?" I asked.

"Only rarely," he replied. "The other thing is that we don't know if it works. In terms of Chinese medicine, there are thousands and thousands of preparations whose efficacies have never been tested empirically. Nowadays there's a rule of thumb, which I don't agree with, that if the Chinese use it in herbal medicine, it must work. That's really a silly attitude. But I wouldn't call the Chinese a placentophagic society."

On a more recent and Western note, I had also come across a report that in rural Poland in the mid-20th century, peasants "dry [placenta] and use it in powdered form as medicine, or the dried cord may be saved and given to the child when he goes to school for the first time, to make him a good scholar." I ran this by my Polish bat biologist colleague Wieslaw Bogdanowicz, who did a bit of investigation himself (presumably asking some of the peasants he knew if they'd heard of such a practice). The answer came back "*nie*," with my friend telling me it was probably safe to assume that the iPad had overtaken the uCord as an educational tool.

"So why is placenta-eating becoming more popular in the U.S.?" I asked Kristal, mentioning the flurry of recent articles, including the subtle classic, "I Ate My Wife's Placenta Raw in a Smoothie and Cooked in a Taco."

"There were two trends," he told me, "one in the sixties and seventies, and one now. The first had a lot to do with a kind of back-to-nature-hippie-commune philosophy. You'd hear these media

---

reportedly engaging in turf battles with medical personnel, some hospitals banned doulas while others started internal doula programs—presumably in an effort to reduce the number of birthing-room-related fistfights.

reports based on anecdotal evidence that at some commune, when one of the members gave birth, they took the placenta and cooked it up in a stew and everybody partook of it. It wasn't linked to medical benefits, really."

Kristal continued. "What's happening now with this fad is that midwives and doulas are responsible for spreading the word that ingesting placenta after delivery has positive female health benefits. The evidence, when you try to track it down, is anecdotal, spread by [these people] and by their journals." Significantly, Kristal mentioned that there hadn't been a single double-blind, placebo-controlled study on the reputed benefits to humans of placentophagy.

To learn why people were currently eating their own placentas, I contacted Claire Rembis, the owner/founder of Your Placenta, a one-stop center for all of your placenta-related needs. Based out of her home in Plano, Texas, Rembis not only offers the standard placenta encapsulation services but will also prepare placenta skin salves, placenta-infused oils, and placenta tinctures, which she described as "organic alcohol" in which a mom's placenta has been soaked for six to seven weeks. Additionally, there's placenta artwork, in which a client's placenta can be used to make an impression print (balloons and flowers seem popular, with the umbilical cord standing in quite nicely for the balloon string or flower stem). During the impression-making process, vegetable- and fruit-based paints are dabbed on to the placenta, which is then pressed between a clean surface (like a diaper changing pad) and a piece of heavy art-stock paper. Immediately after its modeling gig, the placenta is rinsed off and undergoes further preparation in order to assume its rightful place within a gel capsule. For those moms who want to keep their placenta closer to their heart, Claire also makes necklaces—tiny, stoppered bottles, full of "placenta beads" (a secret formula) and available with or without gemstones.

Soon after an introductory email, Claire invited me down to Dallas. I thanked her but declined, explaining that I'd just begun a new school year at LIU.

"Well, if you're ever in Texas, I'd be happy to prepare one for you," she responded. "I browsed your website and think you might enjoy it."

*Wait a minute*, I thought as I read her email. *Was she inviting me down to Texas to eat placenta?*

I followed up.

She was.

"I've got some of my daughter's placenta in the freezer right now," Claire said. She mentioned that her husband, William, was a chef and loved coming up with new placenta-based recipes. "A few months ago I worked with a mom who wasn't keen on swallowing pills," Claire told me. "So, we made her placenta tea and chocolate placenta truffles."

With an offer like that on the table, what else could I do but book a flight to Dallas?

TWO NIGHTS BEFORE my trip to Texas, I received an email from Claire. "We have a bit of an adventure heading our way this Friday," she wrote. Apparently her babysitter had cancelled at the last minute, "and so our sweet little angels will be here for the placenta fun."

*No problem*, I thought and kept reading. I soon learned that not only were her kids "amazing," they were homeschooled, and there were ten of them.

"Well, this is definitely going to make things more interesting," I wrote back.

As I flew into Dallas the evening before our meeting, the pilot of our 737 skirted an enormous weather front, providing me with a spectacular view of the light show unfolding within the towering wall of clouds. We landed less than an hour after 85-mile-per-hour, straight-line winds had strafed the Big D and its suburbs, doing a serious number on the electrical grid and (as I would learn) temporarily knocking out the power in Claire's house. There was definitely an unsettling vibe in Dallas and it wasn't just the weather. Two days earlier, America's first Ebola patient, a Liberian named Thomas Duncan, had fallen ill at an apartment complex not far from the hotel where I was staying, and now over a hundred people who'd come into contact with the sick man were being alerted—some of them quarantined.[42]

I pulled up to the Rembis house a little after 6 PM with a bagful of camera gear and a bottle of Amarone. Surprisingly, the clerk at a local liquor store had no idea what wine would go well with placenta. I went with an Italian red. Claire's husband, William, had previously narrowed my menu choices down to "placenta fajitas with hatch pepper and cilantro rice" or "placenta *osso buco* with sides." Considering my ethnic background, I opted for *cannibale italiano*.

Seconds after chef William ushered me into their ranch-style home in a quiet suburban neighborhood, I was literally hit by a wall of children—touching my hair and shoulders ("soft" and "higher than Dad's," respectively), asking me if I had an iPhone 6 ("Yes"), asking if they could hold my iPhone 6 ("Maybe later.").

---

42 Mr. Duncan died about a week later, his case igniting a media-fanned fear fest reminiscent of the early days of the AIDS epidemic.

I flashed back to a scene in one of my favorite films, *Raising Arizona*, as Nick Cage's character describes his first encounter with the Arizona Quintuplets to his wife, played by Holly Hunter. "They started crying and they were all over me. It was kind of horrifying, honey."

For a few seconds I actually thought about bolting.

Noticing my rather obvious distress, William Rembis, an amiable, 40-something-ish longhair, broke out his own bottle of wine and handed me a Dixie cup. I quickly gulped down the contents.

As the bell sounded for Round 2, most of the Rembis kids now had something to show me ("Look how far I can stretch my ear"), tell me ("We're going to watch *Babe* now"), or ask me ("Are you going to eat my mom's placenta?"). Unlike the fictional Arizona quints, the Rembis children came in a variety of sizes—from a peapod infant up to a husky 15-year-old boy. There seemed to be about a half dozen cute little girls including a pair of 8-year-old twins. Two of the kids were named Cinderella and Belle.[43]

Soon enough, I met Claire Rembis, who hadn't been feeling well that day. She had a shy smile, which she flashed only rarely. Almost immediately Claire sent William scurrying off to Wal-Mart for some placenta prep supplies: disposable diaper changing pads, paper towels, and fresh garlic.

I decided to interview Claire on her front lawn and she politely told her kids to remain inside, the older ones charged with keeping the little guys occupied with something other than their mom. They did so without complaint (or nearly so).

Once we were outside and talking one-on-one, Claire seemed

---

43 In May 2015, I learned that Claire had given birth to her 11th child (a son).

to relax. Throughout the interview, she answered carefully, once or twice asking me to repeat or clarify a specific question.

"What was it that got you interested in consuming your own placenta?" I asked her.

"Hearing about the experiences of other moms from the home-birth midwives that I began working with when I had child number seven. My midwife, who's been practicing since the seventies, explained to me that the placenta was one of things she uses to help moms with problems like post-birth hemorrhaging. Since [midwives] can't prescribe medicines like a doctor, they can only use natural remedies to help moms when they have issues at home. So it was with her encouragement that I decided to try it myself."

I then asked Claire what specific health benefits she thought she was getting from consuming placenta. She responded by first telling me that she'd initiated her own research study, to investigate just such questions. So far Claire had interviewed more than 200 moms, but chose to speak about only those benefits that she herself had experienced. She also said that because she hadn't started consuming her placentas until after child number seven, she clearly felt that she'd established a baseline against which to compare her own placenta-related experiences. Claire explained that after each of her first six births she'd gone through "the baby blues," which she attributed to the "hormonal drop" caused by the loss of her placenta.

"The first thing I noticed after taking placenta products [capsules] after baby number seven [not her real name] was the energy. I felt very energetic. The most significant thing, though, was not feeling like I was on an emotional yo-yo—one minute crying, the next minute happy. Any mom knows exactly what I'm talking about, and it was the thing I dreaded most about having children. Consuming my placenta made me feel a little bit more normal—like I did when I was pregnant, but before giving birth."

Claire went on to tell me that what really convinced her that there were benefits was the fact that she'd get "emotional and out of sorts, and weepy and cranky, when it was time for another pill." When she took one, she said, those emotions leveled out.

"With my eighth child I was severely anemic, both while pregnant and post-partum." She recounted that her hemoglobin counts were low (6 gm/dL), and that rather than taking the iron pills typically described by doctors, she took her placenta pills instead (which she claimed were "packed with iron and all sorts of amazing things"). She also pointed out that she prepared her pills from raw placenta, "so that you're not losing so many of the nutrients" from the cooking process. "By two weeks postpartum, my hemoglobin was just below 10 [gm/dL]." Normal values for an adult woman are 12 to 16. "So not only did I experience the energy and mood stabilization, but I knew that the claims I'd read about, promoting placenta as an iron supplement, were true."

Next, I asked Claire if there was any evidence for the positive benefits she'd spoken about. She immediately brought up Mark Kristal's papers. "Granted," she said, "his research is mostly with mice [rats, actually]. There is, you know, no professional, scientific research in humans right now. Not truly scientific research.[44] What

---

44 In a 1954 study, Czech researchers claimed that placenta consumption increased lactation in postpartum women having lactational difficulty (compared to a control group fed beef). According to Mark Kristal though, "This study does not conform to modern-day ideas about scientific methods or statistical analyses." He noted that "the experiment was methodologically flawed" and that the hormones responsible for increased lactation would have been denatured in the preparation they described.

I'm doing [gathering information on placentophagy] isn't scientific. I'm interviewing and getting feedback from moms. Plus I have my own experiences. Then there's what I hear from midwives who have been practicing for years and years."

In Claire's view, this list was certainly an acceptable alternative to the evidence a more formal scientific study might provide. "To tell you the truth," she went on, "I wasn't completely sold on it [the benefits of consuming placentas] until I actually tried it myself."

I followed up. "Given that some dangerous material has been identified in placental tissue, do you really think it's safe to eat?"

At this point, Claire's description of the placenta as "a filter but not in the sense of a coffee filter" strayed a bit from what I'd learned about the organ. And although the placenta clearly wasn't a coffee filter, she seemed surprised to learn that researchers had determined that it did retain some of the toxic substances and pathogens that it filtered. I told her that I had read about studies in which placental tissue from infected moms also contained hepatitis-, herpes-, and AIDS-infected cells. In response, she agreed that under certain circumstances, consuming the placenta was *not* a good idea. She told me that to avoid coming into contact with pathogens her contract had a clause stating that clients were unaware of having any bloodborne diseases.

I posed the same question to Claire as I had to Mark Kristal. Why did she think there was currently so much interest in placentophagy?

"It just starts with one person trying it," she said. "They see it helps and then they tell another person and so on." She also attributes it to a rise in the popularity of home births, which she said, "have increased dramatically over the past several years."

"People try it and it works for them. Then they tell their friends.

It's just spreading like a virus." Given the local current events, I'm sure I winced at the comparison.

In short order, William and his son Andrew returned with the supplies, and so we headed inside and into the kitchen. Team Placenta quickly split their organ-related duties—he dicing veggies near the stove, she disinfecting the sink-side counter before covering the surfaces and adjacent floor with the aforementioned diaper changing pads. Once the place had been sanitized and covered in absorbent blue, Claire carried over a medium-sized Tupperware container. Prying off the lid, she revealed a roughly Frisbee-shaped organ that was perhaps seven inches across and half an inch thick. (It was smaller than what I was expecting.)

"You won't be eating this one," Claire told me, since it belonged to a client. She gestured to a bed of ice in the sink that held up a small baggie containing what looked to be several small strips of calves' liver. "That's mine," she said.

"And I'll be cooking it up for you," William chimed in happily. Now clad in an embroidered chef's apron, he was chopping away at carrots and tomatoes. "All organic," he assured me (and thank goodness for that).

Wearing disposable gloves, Claire placed her client's placenta on a pad, unfolded it a bit, and allowed me to move in for a peek. The surface was irregularly shaped and reminded me more of scrambled eggs than an organ (albeit liver-colored scrambled eggs holding clots of bluish blood).

"This side faced the wall of the uterus," Claire told me as she de-clotted the irregular surface. She spent several more minutes examining the placenta carefully (seemingly looking for defects) before gently flipping it over like a large bloody pancake. This side was smooth, dark blue, and glistening. A fan of large blood vessels ran from the periphery, converging on a 12-inch section of umbilical cord and winding around it like the stripes on a barber's pole.

I turned my attention to William, who was sweating vegetables in a sauté pan. He added a little beef stock, allowing the flavors of the tomatoes, garlic, and onions to mingle as the veggies softened. A minute or two later he retrieved the baggie containing his wife's placenta from its ice bath and emptied the bloody slivers onto a paper plate. As I watched (*it still looked like liver*), the chef scraped the meat into the pan. Within seconds the kitchen was filled with an aroma that reminded me of beef.

The thin strips coiled up during the cooking process, now looking a bit like larger versions of the bacon chunks in a can of pork and beans, but without the fat. William added about a quarter cup of the Amarone—the steam rising up as the placenta simmered.

It smelled delicious.

Two or three minutes later, William plated my placenta *osso buco* and passed me the dish. Without hesitation, I took a forkful—making sure to skewer two of the four bite-sized pieces. Placing Claire Rembis's placenta into my mouth, I started chewing.

Before experiencing placentophagy firsthand, I had done some research into what human flesh might taste like. I was somewhat puzzled at the scarcity of credible reports, although a number of notable cannibal crazies had been perfectly happy to discuss the topic.

The term "long pig" has become the most popular reference point to describe the supposed porklike taste of human flesh. The oldest reference I could find comes from a letter written by Rev. John Watsford in 1847, describing the practice of ritual cannibalism practiced by the inhabitants of the Marquesas Islands, a group of approximately 15 Polynesian islands located around 850 miles northeast of Tahiti. But while the letter does represent the translation of a Polynesian term for the use of human flesh as food, there is no real mention of how it tasted.

> The Somosomo people were fed with human flesh during their stay at Bau [a tiny Fijian islet], they being on a visit at that time; and some of the Chiefs of other towns, when bringing their food, carried a cooked human being on one shoulder, and a pig on the other; but they always preferred the "long pig," as they call a man when baked.

More reliable support for the pork hypothesis came from the infamous cannibal Armin Meiwes, who is currently serving a life term for killing and devouring Bernd Brandes. The latter, a 42-year-old computer technician, answered Meiwes's cannibalism chat room post in 2001. It was the perfect match, with Meiwes obsessed with cannibalism and Brandes fixated on being eaten. Shortly after entering Meiwes's dilapidated house in Rotenburg, the new friends decided to sever Brandes's penis, which they reportedly tried to eat raw. Finding it too tough and chewy, they set out to cook the schnitzel but overcooked it—Meiwes eventually feeding it to his dog. Brandes, nearly unconscious from a combination of blood loss and the pills and alcohol he'd swallowed, eventually died—helped along by the knife-wielding Meiwes. The Internet's first cannibal killer then dismembered his suddenly former pal. He stored the body parts in a freezer and consumed them over the course of several months.

"I sautéed the steak of Bernd with salt, pepper, garlic, and nutmeg," Meiwes told interviewer Günter Stampf. Reportedly Meiwes ate more than 40 pounds of Mr. Brandes during the months following the killing. "The flesh tastes like pork, a little more bitter," he said, noting that that most people wouldn't have been able to tell the difference. "It tastes quite good."

The pork comparison, however, was not shared by all.

Issei Sagawa, an unrepentant Japanese cannibal, who murdered and ate a female Dutch student in 1981 (and got away with it because of powerful family connections), compared his victim's flesh to raw tuna.

While we're on the topic of Meiwes and Sagawa (albeit briefly), some readers may be wondering why I've essentially steered clear of the criminal cannibalism typified by this pair and their ilk. One reason is that the topic has been covered in sensational (and often gory) detail in a number of previous books. More importantly, though, several of these psychopaths are still alive (or recently deceased) and out of respect for the families and loved ones of their victims, I have chosen not to provide these murderers with anything that could even vaguely be interpreted as acclaim.

In the 1920s, *New York Times* reporter William Seabrook set out to eat a chunk of human rump roast with some Guero tribesmen in West Africa. Upon returning home he began writing a book about his adventures. Depending on what source you believe, either Seabrook discovered that the tribesmen had tricked him into eating a piece of ape, or they had simply refused to share their meal with him. With the validity of his book in jeopardy, Seabrook set out to procure some real human flesh—this he claimed to have gotten from an orderly in a Paris hospital who had access to freshly dead patients. Seabrook says that he cooked the meat over a spit—seasoning it with salt and pepper and accompanying it with side of rice and a bottle of wine. It did not taste like pork, he said, "It was good, fully developed veal, not young, but not yet beef."

Back in Plano, Texas, the Rembis family stood by waiting for my reaction, I took my time, chewing Claire's placenta slowly. The first thing that came to mind wasn't the taste—it was the texture. Firm but tender, it was easy to chew.

*The consistency was like veal.*

The taste, though, had none of the delicate, subtly beefy flavor of veal. *Definitely dark meat—organ meat*, I thought, but it wasn't exactly like anything I'd ever eaten before. It had a strong but not overpowering flavor. I swallowed Claire's placenta and picked up another forkful.

It tasted very much like the chicken gizzards we'd fried up as college students. "It's very good," I told the assembled Rembis clan and they responded with a chorus of moans, groans, and giggles.

A few minutes later, I had cleaned my plate.

As expected, the Rembis kids were full of questions.

"Can I hold your iPhone now?"

"Can we ask Siri a question?"

"Do you want one of my Pringles?"

I squatted down to kid height, pulled out my phone and got on with the important stuff, hoping to avoid one of the of the side effects of placenta eating I'd read about: unpleasant burps.

• • •

So IS THERE any real benefit to the practice of placentophagy? If one were to gauge the benefits by the number of societies that engage in it, the answer would be a resounding "Nope."

Maggie Blott, a spokeswoman for the Royal College of Obstetricians and Gynaecologists (UK), believes that there's no medical justification for humans to consume their own placenta. "Animals eat their placenta to get nutrition—but when people are already well-nourished, there is no benefit; there is no reason to do it."

But what about the alternative scenario—that consuming placentas could possibly have detrimental effects?

According to Mark Kristal, "The sharp distinction between the prevalence of placentophagy in non-human, non-aquatic mammals, and the total absence of it in human cultures, suggest that different mechanisms are involved. That either placentophagia became somehow disadvantageous to humans because of illness or sickness or negative side effects, or something more important has come along to replace it."

Ultimately, though, the possibility of negative effects and the lack of evidence for beneficial effects doesn't faze folks like Claire and William Rembis and, similarly, it didn't prevent Oregon representative Alissa Keny-Guyer from sponsoring bill HB 2612, which was passed unanimously by the state Senate in 2013. The new law allows Oregon mothers who have just given birth to bring home a second, though slightly less joyous, bundle when they leave the hospital.

Except in rare cases, it appears that medicinal cannibalism is at worst a harmless placebo. But, if that's true, then beyond our culturally imposed taboo, maybe there exists another reason why we don't indulge in cannibalism on a more regular basis. Recalling that UNLV researchers found no mention of placentophagy in the 179 societies they examined, I wondered if perhaps these groups knew something that ritual cannibals, proponents of medicinal cannibalism, and modern placentophiles have missed.

# 17: Cannibalism in the Pacific Islands

*Nothing it seems to me is more difficult than to explain to a cannibal why he should give up human flesh. He immediately asks, "Why mustn't I eat it?" And I have never yet been able to find an answer to that question beyond the somewhat unsatisfactory one, "Because you mustn't." However, though logically unconvincing, this reply, when backed by the presence of the police and by vague threats about the Government, is generally effective in a much shorter time than one could reasonably anticipate.*

—J. H. P. Murray, Lt. Governor/Chief Judicial Officer, British New Guinea, 1912

In the spring of 1985, veterinarians working in the English counties of Sussex and Kent were puzzled when dairy farmers reported that a few of their cows were exhibiting some peculiar symptoms. The normally docile creatures were acting skittish and aggressive. They also exhibited abnormal posture, difficulty standing up and walking, and a general lack of coordination. Most of the cows were put down and sent on to rendering plants—facilities that process dead, often diseased animals into products like grease, tallow, and bone meal. It wasn't until the following year that

England's Ministry of Agriculture, Fisheries and Food launched an investigation.

According to research biochemist Colm Kelleher, microscope slides were prepared from the brains of stricken cows and they showed the tissue to be riddled with holes, reminiscent of Swiss cheese. In what would become the first of many unfortunate decisions, the veterinary pathologists who examined the slides blamed the holes on faulty slide preparation. But by November of the following year, researchers knew that the abnormal spaces had once been filled with neurons that had shrunken and died. They also thought that amyloid plaques, the sticky concentrations of a normally non-sticky brain protein, might be a contributing factor to the neuron deaths. The holes and plaques were characteristic of a number of neurological diseases, with sheep scrapie and Creutzfeldt-Jakob Disease (CJD) being the best known of these somewhat mysterious maladies. These and other diseases of their ilk were classified as transmissible spongiform encephalopathies (TSEs) because of the spongy appearance of infected brain tissue.[45] The British researchers soon named their new disease Bovine Spongiform Encephalopathy (BSE). The press, of course, would need something a bit splashier. They settled on "Mad Cow Disease."

By 1987, there were over 400 confirmed cases of BSE, which had spread to cattle herds across England, and while scientists looked for puzzle pieces, nervous government officials (who preferred the term "bovine scrapie") repeatedly reassured the public that it was safe to eat English beef. And why not, they rationalized, hadn't scrapie been killing sheep for centuries with no harm to the humans

45 Admittedly, "Swiss cheesiform" doesn't have the same ring.

who consumed them? Why then should a bovine version of the disease be any different?

Other researchers, though, were not so sure, and a few of them began comparing BSE to a disease that *had* killed humans—thousands of them. To these professionals, this particular affliction was still known by its indigenous name, kuru (the trembling disease). Like their UK counterparts, the American media corps had previously scrambled to coin their own inappropriate names for kuru. They settled on "the laughing death" and alternately, "laughing sickness," at a time when being called "politically correct" meant you had voted for the guy who won.

According to the rash of radio, magazine, and newspaper accounts that followed the discovery of kuru by Western researchers in the 1950s, some of the earliest and most unsettling symptoms of the disease were intermittent bouts of uncontrolled laughter, or what researchers referred to as pathological laughter. But if the symptoms of kuru were disturbing to the mid-20th-century public, the way the disease was said to spread was even worse, for according to those working to unlock the lethal mystery, kuru was spread by cannibalism.

In the early 1950s, anthropologists and medical researchers began arriving at one of the wildest and most primitive regions on the planet. New Guinea, the world's second largest island, rises from the western Pacific like a dinosaur with a mountainous spinal column. Upon their arrival, the scientists saw no roads and nothing much that resembled their concept of a city or even a town. Instead they found themselves crossing parasite-ridden mangrove swamps and rainforests whose primary inhabitants seemed to be biting insects, terrestrial leeches, and venomous snakes. But even after reaching their destination in the foreboding Eastern Highlands, conditions were no less dangerous, for the researchers had come to study New Guinea's infamous cannibals.

Numbering approximately 36,000 individuals in the mid-20th century, the Fore (pronounced FOR-ay) spoke three distinct dialects and inhabited some 170 villages situated among New Guinea's lush mountain valleys. Desiring (and having) little or no contact with the outside world, the Fore practiced the same slash-and-burn agriculture that had sustained them for thousands of years. Currently what made them especially interesting to the researchers was not their lack of contact with the modern world, their farming techniques, or even the reports that they practiced ritual cannibalism. It was the fact that something was killing them—horribly and at an alarmingly rapid rate.

A decade earlier, as post-WWII colonialism extended its reach onto the "primitive" Pacific islands, Australian patrol officers in New Guinea began encountering some of the most isolated of the island's inhabitants.[46] Like the missionaries that had arrived, preached, and often disappeared for their troubles, the Australian officials (whom the Fore called *kiaps*) encouraged the locals to curtail what they considered some unacceptably bad behavior. Sorcery and tribal warfare, the Aussie officials said, were prohibited. The Fore were also requested to please stop eating each other, which they claimed to do as a way of honoring their dead. Grudgingly, the indigenous people soon agreed to the requests, although today

---

46 The Territory of New Guinea was administered by Australia from 1920 until 1975.

many anthropologists believe that, while most of them complied, others simply concealed their long-held rituals whenever the nosy white people came around.

Unfortunately, many of the Fore were experiencing problems far more serious than the arrival of their pushy new friends. In fact the *kiaps* were reporting that something akin to a plague was taking place, one that primarily took its toll on women and children. In addition to the uncontrollable laughter, victims experienced tremors, muscle jerks, and coordination problems that gradually gave way to an inability to swallow, and finally, complete loss of bodily control. The Fore responded to their stricken relatives with kindness—feeding, moving, and cleaning them when they could no longer care for themselves. Invariably, though, their loved ones died, all of them—of starvation, thirst, or pneumonia, their bodies covered in bedsores. The mystery disease was killing approximately 1 percent of the population each year.

Fore elders told the foreigners that the sickness resulted from a form of sorcery. The *kiaps* were informed that the process went something like this: Sorcerers would stealthily obtain an item connected to their intended victim, like feces, hair, or discarded food. After wrapping the object in leaves, they would place it in a swampy area where it couldn't be found. As the sorcery bundle began to decompose, so, too, the Fore said, would the victim. The elders also told the *kiaps* that the condition could not be cured or even treated, and they tried to explain that preventing this sort of thing was the main reason they sometimes killed each another. The patrol officers took it all down, and although some of them were quite sympathetic, they drew the line at allowing the Fore to resume the killing of suspected sorcerers. These unfortunates were generally men or boys accused without evidence (usually several days after someone in their own village had died of kuru), then hacked, stoned, or bludgeoned to death in a form of ritual murder known as *tukabu*.

It made sense to the *kiaps* that the best way to stop the killing was to gain an understanding of the mystery ailment, and a number of hypotheses were developed. Initially, it was suggested that the deaths were stress-related and had resulted from the Fore making contact with the whites. Others, though, feared that whatever was killing the Fore had a more pathological nature.

By the time Ronald and Catherine Berndt arrived in the New Guinea highlands in 1951, they had already spent years in the field studying Australia's aboriginal communities. The thought was that the Fore would become another notch on Ronald's impressive anthropological belt and, at first, things looked promising. The indigenous highlanders threw parties for the couple, reportedly believing them to be the spirits of their dead ancestors, returning to the fold to relearn the language they had apparently forgotten. Soon enough, though, the Fore lost interest in rehabilitating their pale relatives—but not in the strange goods they had brought along with them. Fascination soon turned to envy, and not long after the Berndts settled in, they wrote that the locals were "difficult people to deal with," requiring "payments for stories: salt, tobacco, newspapers, wool strands, matches, razors, and so on." The anthropologists also reported "plenty of cannibalism."

"Actually these people are 'bestial' in many ways," Ronald Berndt wrote. "Dead human flesh, to these people is food, or potential food." He also described cannibalism among the Fore as an outlet for sexual violence, and "orgiastic feast" was a phrase he seemed to regard with fondness.

A decade later, the not-yet-controversial anthropologist Bill Arens commented on Ronald Berndt's influential 1962 book on social interactions among the Fore. According to Arens, Berndt's tome, *Excess and Restraint*, displayed "too much of the former and too little of the latter." Arens was particularly galled by Berndt's description of a Fore husband copulating with a corpse as the man's wife simultaneously butchered the body for a meal. As these things go, she accidently cut off her husband's penis with her knife. "Now you have cut off my penis!" the man cried. "What shall I do?" In response, according to Berndt, the woman "popped it into her mouth, and ate it. . . ."

Arens was not alone in his criticism of the Berndts, as others concluded that, while the pair had made some important anthropological contributions, there were more than a few problems with their work. Most of these related to the many instances of outrageous behavior Berndt detailed in his book—coupled with the growing suspicion that he had made much of it up.

As I began my own research on cannibalism, I found it odd that the Berndts' reputation, especially as it related to their claims about extreme behavior by the Fore, seemed to have recovered quite nicely with the passage of time. In fact, I noticed that the Berndts were cited in many of the more recent papers on kuru and cannibalism as having presented solid evidence that the Fore practiced man-eating.

But more on that later.

Back in the New Guinea Highlands of 1952, relations between

the Berndts and the Fore failed to improve during the couple's second field season. Ronald reportedly slept with a pistol under his pillow, at one point firing it to scatter some villagers who had been bothering him. As for kuru, the Berndts had seen the disease among the Fore but apparently never made any meaningful connections concerning its cause. Within two years of their arrival in New Guinea, the pair left the country for good.

Fortunately, after the Berndts' inauspicious start, the researchers who followed them had far better luck, some of them initiating studies that would become classics in the fields of anthropology and disease hunting—and eventually garnering a pair of Nobel Prizes.

One of the researchers was Daniel Carleton Gajdusek (GUY-doo-shek), a Yonkers, New York, native and 1946 graduate of Harvard Medical School. Gajdusek had no real interest in practicing medicine but instead chased his fascination with viral genetics and the anthropology of what he called "primitive" communities across the world. He studied rabies and plague in the Middle East, hemorrhagic fever in Korea, and encephalitis in the Soviet Union. Arriving in Melbourne in 1955, the brilliant but eccentric researcher frequently "went bush," studying child development among the aboriginals and collecting blood serum for several Australian research labs.

Gajdusek flew to New Guinea in 1957 and, with nothing but his own meager funds to support this venture, he began working on a new problem. To a colleague in the United States, Gajdusek wrote:

> I am in one of the most remote, recently opened regions of New Guinea, in the center of tribal groups of cannibals, only contacted

in the last ten years—still spearing each other as of a few days ago, and cooking and feeding the children the body of a kuru case, the disease I am studying—only a few weeks ago.

But Gajdusek had never seen any actual cannibalism and he had very little real knowledge about kuru. Beyond the stress-related hypothesis, there was some conjecture that the deadly condition might be the result of an environmental toxin. Others believed that kuru was a hereditary disorder. Consequently, Gajdusek got busy. He spent months collecting blood, feces, and urine from the locals. He ran tests on those stricken by the disease and, with the aid of translators, he conducted interviews with victims and their family members.

By mid-1957, Gajdusek was working with Vin Zigas, a medical doctor who had already been gathering information, as well as his own blood samples. That November their initial findings were published in the prestigious *New England Journal of Medicine*. Kuru, the authors claimed, was a degenerative disease of the central nervous system. They described the clinical course of the disease as well as its curious preference for striking three times as many women as men. The skewed sex ratios were difficult to pick up, however, since more men were being killed for having been kuru sorcerers. For the Fore, ritual murder had become the great equalizer. In what would later become an important observation, Gajdusek also noted that kuru occurred equally in children of both sexes.

The researchers sent off blood and tissue samples for analysis but the results showed no evidence of viral infection, nor did they appear to indicate the presence of a toxin (they had suspected that the Fore were possibly being poisoned by heavy metals in their diet). But a number of the tissue specimens did show something remarkable—although it was as much about what the samples lacked as what they exhibited. After examining the brains of eight kuru victims, scientists at the National Institute of Health (NIH) in Bethesda, Maryland, reportedly found no evidence of inflammation or any immune response. That meant the victim's body had not been fighting off a pathogenic organism like a virus, bacterium, or fungus. In most cases, at the first signs of a viral, bacterial, or fungal intruder, the body initiates a sustained, multi-pronged defense consisting of non-specific responses like swelling and inflammation, and cell-mediated responses like the production of antibodies—proteinaceous particles that bind to foreign proteins (antigens) found on foreign cells and viruses.[47]

What the investigators did find was that large regions of the cerebellum (which sits like a small head of cauliflower behind the cerebral hemispheres) were full of holes.

Igor Klatzo, an NIH researcher, had seen a disease like this only once before. "The closest condition I can think of," he said, "is that described by Jakob and Creutzfeldt."

Another NIH scientist noticed a similarity between kuru and the transmissible spongiform encephalopathy (TSE) known as scrapie, an unusual infectious agent of sheep. Scrapie, which was present in European sheep by the beginning of the 18th century, was named for one of its symptoms, namely the compulsive scraping of the stricken animal's fleece against objects like fences or rocks. It had

47 Once attached to their specific antigens, the antibodies either interrupt the normal function of the foreign cell or virus, or mark it for destruction by other cells of the immune system like macrophages.

been previously been classified as a "slow virus," an archaic term for a virus with a long incubation period, in which the first symptoms might not appear for months or even years after exposure. Klatzo and William Hadlow, who had made the kuru/scrapie connection, now suspected that the cause of kuru might also be a slow virus.

At this point, Ronald Berndt tried to reassert himself as the world's leading authority on the New Guinea highlands and its indigenous inhabitants. Miffed that medical researchers were intruding on what he considered his anthropological turf, he refused to be outdone by upstarts like Gajdusek. Berndt wrote his own article on kuru, reemphasizing the importance of sorcery and resurrecting the original stress-related explanation. Fear alone, Berndt claimed, was probably enough to initiate the irreversible symptoms of kuru.

Fortunately, though, in what many might argue was a strong positive step for kuru research, Berndt and his wife refused to share their "data" and "expertise" with those who were actually doing real research on the disease. Gajdusek, for his part, dismissed Berndt's baseless assertions, believing instead that the high occurrence of kuru among young children argued against a psychological origin for the disease. He was leaning toward the explanation proposed by genetics professor Henry Bennett, who attempted to explain the discrepancy between male and female adults dying of kuru.

Bennett proposed that a mutant kuru gene "K" was dominant (K) in females but recessive (k) in males. Accordingly, only males who were KK (and who had inherited a dominant form of the gene from each parent) died of kuru, while males who were either normal (kk) or carriers (Kk) were unaffected by the disease. Alternately, females who were either KK or Kk died of kuru, while only those females who were normal (kk) were unaffected.

In the end, the fact that kuru victims included equal numbers of male and female children, but few adult males, was deeply troubling to Gajdusek, and it raised serious questions about Bennett's gene-based disease hypothesis, which was soon abandoned.

By this time the researchers had already been dealing with another problem—this one related to the sensationalized slant the press had given the kuru story. *Time* magazine, for example, opened its November 11, 1957, article "The Laughing Death" with the following:

> In the eastern highlands of New Guinea, sudden bursts of maniacal laughter shrilled through the walls of many a circular, windowless grass hut, echoing through the surrounding jungle. Sometimes, instead of the roaring laughter, there might be a fit of giggling. When a tribesman looked into such a hut, he saw no cause for merriment. The laugher was lying ill, exhausted by his guffaws, his face now an expressionless mask. He had no idea that he had laughed, let alone why. . . . It was kuru, the laughing death, a creeping horror hitherto unknown to medicine.

The *Time* story then went on to describe how the Fore were "only now emerging from the Stone Age" and that they still practiced cannibalism and the ritual murder of kuru sorcerers ("when they think they can get away with it"). Even when dealing with the scientific aspects of the story, the article's anonymous author took ghoulish satisfaction in reporting Gajdusek performed autopsies without gloves, atop the same dining-room table where meals were eaten. Additionally, the article continued, the researcher, "had to haggle with victims' relatives for the bodies" of kuru victims and "he got some bodies at the bargain price of only one ax." For his part, Gajdusek hated the media coverage and he considered the term "laughing death" to be a "ludicrous misnomer."

The worldwide media coverage did have at least one positive effect, in that it increased the public's awareness of the deadly problem facing the Fore. Because of this, universities began to funnel funds into kuru research, and this money helped support a new influx of professional researchers into the region.

Two of the first to arrive were cultural anthropologists Robert and Shirley Glasse (now Shirley Lindenbaum), who came from Australia to New Guinea on a university grant in 1961. Studying kinship among the Fore, they returned to continue their research in 1962 and 1963. Their work in the New Guinea Highlands would ultimately allow them to make the connection between kuru and cannibalism.

I MET DR. Shirley Lindenbaum half a century later at an Upper West Side apartment, which had been decorated with art and other memorabilia collected during a long and distinguished career. In a voice that still retained the hint of an Australian accent, she talked about her studies.

"What was it that finally convinced you that cannibalism was the mode of kuru transmission?" I asked her.

Lindenbaum explained that once the epidemic began in the New Guinea Highlands, she and her husband were instructed to gather genealogical data about people who had kuru. In doing so, they spoke to Fore elders who had seen the first cases of the disease in their villages.

"They could remember these cases and even the names of the people in the North Fore who came down with the disease some couple of decades earlier. There were these tremendously convincing first stories and we said, 'What happened to those people?' And the Fore said, 'Well, they were consumed.' We *knew* they were cannibals."

I pressed on, asking Lindenbaum how she knew for sure that the Fore were cannibals. Without hesitation, she cited "fieldwork in the area by Ronald and Catherine Berndt in the 1950s" as well as "government patrol reports throughout the Eastern Highlands."

I may or may not have raised an eyebrow at the mention of the Berndts but I did ask Lindenbaum if she was bothered by the fact that no anthropologist, including the Berndts, had ever seen ritualized cannibalism firsthand.

"No," Lindenbaum replied. "Because there are a lot of things we haven't seen firsthand—sexual intercourse among them. But there's evidence that it occurs."

At this point, something like an alarm went off in my brain. Basically, this line about sexual intercourse has become something of a mantra for those anthropologists who claim that ritual cannibalism occurred in a particular group, even though they had not seen it with their own eyes.

After encountering this undeniably catchy analogy several times in the literature, I became curious as to its origin. As far as I can tell, it was coined by Pulitzer Prize–winning anthropologist Jared Diamond, in an article he wrote for *Nature* in 2000.[48] Here's the relevant passage:

> Finally, any society has practices considered acceptable in private but inappropriate to practice in public, in the presence either of

48 For the record, Diamond's stance is that cannibalism was a widespread practice throughout human history.

> anyone else (for example, sex or defecation) or of non-clan members (for example, initiation rites or cannibalism). The abundance of New Guinea babies, my knowledge that babies are conceived by sexual intercourse, and secondhand accounts persuade me that New Guineans practice sex, but I have no firsthand observations of it even after many years there.

The reason for my unease at the mention of this particular notion was based on my interview with Bill Arens, who eventually acknowledged a "clear-cut association" between cannibalism and kuru. Still, though, he had a major problem with the cannibalism/sexual intercourse comparison. Approximately two seconds after asking Arens what he thought about Diamond's famous line, he answered my question with one of his own.

"You ever been in the field with an anthropologist?"

I admitted that I hadn't.

"They're always screwing the natives! So they know sexual intercourse takes place."

"O . . . kay." I said, scanning the paper I had been holding for another question, something about Arens's favorite ice cream flavor, perhaps.

The anthropologist smiled at what I thought was some well-camouflaged discomfort on my part. "Well, I'm just telling you," he said. "You gave me the example. And it's absurd. No anthropologists should ever say that we don't know sexual intercourse takes place among those people because we've never seen it, because that's really a falsehood. They've seen it. And I *know* they've seen it. And if they haven't seen it . . . they should get a mirror!"

Months later on the Upper West Side, Lindenbaum was completely unaware of my unease with what had obviously become a standard line among anthropologists studying aspects of behavior that were, for whatever reason, hard to observe. She continued with her story.

"We knew cannibalism was customary in this area but that the disease had only appeared in the last few decades. And so we thought, *Well, that's very interesting.* When we began collecting ethnographic data about who ate whom, it became clear that it was adult women, not adult men, but children of both sexes. At that time the director of kuru research in New Guinea was a guy named Richard Hornabrook, a neurologist. And he said to us, 'What is it that adult women and children do that adult men don't do?' and we said, 'Cannibalism, of course.' The epidemiological evidence matched the cultural/behavior evidence, and that matched the historical origin evidence. It was such a neat package, you know?"

I nodded. "So what did you do with that information?"

"We told everybody," she said.

"And?"

"And nobody believed us."

Nevertheless, Robert Glasse published his and his then-wife, Shirley's, hypothesis that kuru was transmitted by consuming the body parts of relatives who had died from the disease. As support, he cited the fact that women commonly participated in ritual cannibal feasts but not men. He also wrote that children of both sexes had become infected because they accompanied their mothers to these ceremonies and participated in the consumption of contaminated tissue, including brains. Finally, Glasse calculated that kuru appeared anywhere between four and 24 years after the ingestion of cooked human tissue containing an unidentified pathogenic agent.[49]

Nearly 50 years after Glasse published the couple's findings, anthropologist Jerome Whitfield and his colleagues used an extensive set of interviews as well as previously collected ethnohistorical data to provide a detailed description of Fore mortuary rites. Whitfield told me how his research group deployed the "educated young Fore members of their team to conduct unstructured interviews in a sample group composed of elderly family members who had witnessed, taken part in, or were informed about traditional mortuary feasts."

The interviews revealed funerary practices that ranged from burial in a basket or on a platform to the practice of "transumption," a term Whitfield and some of his colleagues adopted as an alternative to using "cannibalism" to describe the ritual consumption of dead kin. As for how the funerary practices would be carried out: If possible, according to Whitfield, the dying person made his or her preference known. In other cases though, the deceased's family made the call. Generally, the Fore believed that it was better to be consumed by your loved ones than by maggots and, that by eating their dead, relatives could express their grief and love, receive blessings, and insure the passage of the departed to *kwelanandamundi*, the land of the dead. For these reasons, transumption was the funerary practice favored by the Fore.

According to those interviewed by Whitfield's team, the corpse

---

49 We now know that the symptoms may not appear until five decades after exposure.

was placed on a bed of edible leaves in order to ensure that "nothing was lost on the ground as this would have been disrespectful." The body was cut up with a bamboo knife and the parts handled by several women whose specific roles were defined by their relationships to the deceased. Pieces of meat were placed into piles to be divided up among the deceased's kin. Next, the women leading the ceremony enlisted the daughters and daughters-in-law of the deceased to cut the larger pieces of flesh into smaller strips, which were stuffed into bamboo containers with ferns and cooked over a fire. Eventually the deceased's torso was cut open, but during this portion of the ritual the older women formed a wall around the body to prevent younger women and children from seeing the removal of the intestines and genitals. These parts were presented to the widow, if there was one. Once the flesh was cooked, it was scooped out and placed onto communal plates made of leaves. The funerary meal was shared among the dead person's female kin and their children.

The head of the deceased also became part of the ritual. It was cooked over a fire to remove the hair before being de-fleshed with a knife. Next, the skull was cracked with a stone axe and the brain was removed. Considered to be a delicacy, the semi-gelatinous tissue was mixed with ferns, cooked, and consumed. Bones were dried by the fire, which made it easier to grind them into a powder that would be mixed with grass and heated in bamboo tubes. According to the accounts obtained by Whitfield's team, the Fore ate everything, including reproductive organs and feces scraped from the intestines.

Shirley Lindenbaum told me that, initially at least, members of the Fore were receptive about answering questions related to kuru and cannibalism. Later though, "as more missionaries came in and journalists came through and wanted to talk to people, and said 'tell us about that' [i.e., cannibalism] . . . they became very defensive and wouldn't talk to people about it."

So how did kuru spread from village to village and from one region of the Fore territory to another? According to Lindenbaum, kinship relations were the key. She explained that although Fore women moved from their natal homes to marry men from other groups, they still maintained their kinship affiliations with their former communities. When deaths occurred, women from adjacent and nearby hamlets, who were related to the deceased persons, traveled and took part in the mortuary feasts. Similarly, individuals and families who moved into new communities maintained kinship ties with their former communities, especially on special occasions. Additionally, like other diseases throughout history, kuru traveled along well-defined trade/exchange routes, in this case those connecting the villages of the New Guinea Highland.

Factors like these did much to explain how kuru had spread through the villages and additional research put a timeline on the spread. By tracing the path of the kuru reports, from the earliest to the latest, Lindenbaum and her husband calculated that the first cases of kuru occurred around the turn of the 20th century in Uwami, a village in the Northwestern Highlands. By 1920 kuru had spread to the North Fore villages, and by 1930 into the region inhabited by the South Fore.

Jerome Whitfield, who conducted nearly 200 interviews in the kuru-affected region for his dissertation, believes that the practice of cannibalism in the New Guinea Highlands may have begun 40 or 50 years earlier than the first cases of kuru—which would make it sometime in the mid-19th century.

Eventually these findings became strong evidence against a genetic origin for the disease—since had there been a genetic link, researchers would have not have expected the first reports of kuru to begin so suddenly and only 60 years earlier. Additionally, had kuru been a genetic abnormality, in all likelihood it would have reached something known as epidemiological equilibrium, a condition in which the prevalence of a genetic disorder in a population becomes stable, rather than changing over time. In this case, the Glasses' data indicated that, from its first appearance at the turn of the century, kuru-related deaths had increased for the next five decades, peaking in the late 1950s. In 2008, Michael Alpers wrote that kuru deaths among the Fore peaked between 1957 and 1961 with around 1,000 victims. With the prohibition of cannibalism beginning in the late 1950s, the number of kuru deaths hit a steep decline in the 1960s and 1970s, with less than 300 deaths between 1972 and 1976.

The plague was over. Or so it seemed.

# 18: Mad Cows and Englishmen

*Unfortunately, the custom of consuming human flesh, like exotic sexual practices, polygamy, and other alien habits, raises violent, unintellectual passions in the Western scholars who study them.*

—Brian Fagan, *The Aztecs*, 1984

In the 1980s, researchers in the United Kingdom, like those in New Guinea, were also seeking to explain how a strange form of spongiform encephalopathy was being transmitted, and where it had come from. Like their New Guinea counterparts, they struck pay dirt after making a connection to diet—in this case, after examining the diets of English dairy cows.

In order to maximize milk production, farmers typically supplemented livestock diets with protein—most often in the form of soybean products. In England, however, where there is no substantial soybean agriculture, using soy as an additive would have been an expensive proposition. Because of this, in the 1940s meat and dairy industries in the UK began to render the waste products of livestock slaughter into an innocent-sounding material they called "meat and bone meal." Noting the cost-saving benefits, the U.S. and other nations followed suit. In addition to ingredients like bones, brains, spinal cords, heads, hooves, udders, and viscera, the recipe *du jour* for meat and bone meal also called for the bodies of sick animals (including poultry, pigs, sheep, and so-called downer cattle)[50] that had been deemed unfit for human consumption. This gruesome mess was sent off to the "knacker's yard," a British slang term for a rendering plant. During the rendering process, the above-mentioned goodies were ground, cooked, and dried into a greyish, feces-scented powder which was sold as a source of dietary protein, calcium, and vitamins for dairy cows, beef cows, pigs, and poultry.

Although a comparison of livestock-feeding practices with the ritualized consumption of relatives by the Fore seems to be a bit of a stretch, in reality there is an important similarity. In the case of the Fore, ritual cannibalism of kuru victims exposed practitioners to a deadly infective agent. And although nobody knew it at the time, beginning in the 1940s, livestock were exposed to similar pathogens after being forced to consume dietary supplements derived (at least partially) from sickened members of their own species.

But why had the Bovine Spongiform Encephalopathy (or BSE) epidemic struck so suddenly four decades later? The livestock industry had been using meat and bone meal for 40 years and nothing like this had ever happened before.

In searching for answers, the British government enlisted epidemiologist John Wilesmith, who examined the records (where they existed) of rendering plants across the UK. He soon determined that several modifications related to the rendering process had been instituted in the early 1980s. The first was that most of the plants had

---

50 "Downer cattle" is a trade term for cows that have become too sick to walk, or die before being slaughtered.

discontinued the separation of tallow (a creamy fat used to made candles and soap) from the material being converted into meat and bone meal. Previously, dangerous solvents had been used to extract tallow during the rendering process, but after a massive industrial explosion in 1974, safety measures were introduced regarding the handling of solvents in the workplace. Rather than deal with the expensive modifications mandated by the new rendering industry regulations, all but two of the plants chose to abandon the tallow extraction process altogether. As a result, substances that had once been removed by the solvent extraction process now remained in the resulting meat and bone meal. Presumably, these substances included the infective—and still unidentified—agent causing Bovine Spongiform Encephalopathy. There was more, though.

Wilesmith and his team learned from herdsmen that several recent changes had been made to livestock diets. The first was a significant increase (from 1 percent to 12 percent) in the amount of meat and bone meal added to dairy cow feed. Calves were also receiving the protein supplement at an earlier age. As in other spongiform encephalopathies, there appeared to be a direct correlation between the amount of contaminated material ingested and the likelihood of contracting BSE. Similarly, the incubation period for BSE was apparently shorter in younger animals. In theory, then, before industry-wide changes in diet were implemented, calves received less of the contaminated supplement and did not start ingesting it until later in their lives. As a consequence, infected animals would have been slaughtered before they had a chance to get sick.

The results of Wilesmith's epidemiological study were presented to ministry officials in May 1988. He told them that the BSE problem could be traced to the popular nutritional supplement that had been contaminated with sheep scrapie. This material had subsequently been fed to cows, sickening them. In retrospect, Wilesmith's detective work was remarkably efficient, but ultimately his belief that BSE and scrapie were one in the same disease would place British consumers in peril for over a decade.

BACK FROM NEW Guinea in 1963, Carleton Gajdusek realized that his fellow researchers had been correct about the striking similarities between kuru-infected brains, brains from victims of Creutzfeldt-Jakob disease, and those from sheep with scrapie. The puzzle was just beginning to come together when yet another piece was discovered—but this one was several decades old.

In 1947, an outbreak of what would become known as transmissible mink encephalopathy (TME) in farm-raised mink led investigators to search for links between the ranches where infected animals had been identified. They discovered that it was a common practice for adjacently located ranches to share animal feed. In these instances, when mink from one ranch came down with TME, invariably so did animals from the adjacent ranch. The feed itself was a vile mess composed of cereal, fish, meatpacking plant by-products like sheep entrails and other internal organs, and flesh from downer cattle. By the time another outbreak of TME occurred in 1963, veterinary researchers suspected that something very strange had happened—the disease had been transmitted across species, in this case from sheep to mink.

By September 1963, similarities in kuru-, scrapie-, and TME-infected brain tissue, coupled with the discovery that TME and scrapie could be transmitted within and between species, led

Gajdusek and NIH researcher Joe Gibbs to an important experiment. At the Patuxtent, Maryland, lab they inoculated a trio of chimpanzees with liquefied brain tissue from kuru victims. If the chimps came down with the disease, it would prove once and for all that kuru was *not* a genetic abnormality or a stress-related psychosis, but an infectious or transmissible agent. As the antsy Gajdusek left the U.S. for another field season in New Guinea, he worried about the long, symptom-free incubation period for scrapie, which sometimes extended up to five years post-exposure. What if his experimental animals didn't get sick for five years or more?

Gajdusek need not have worried. Less than two years after being inoculated, two of the chimps, Georgette and Daisy, began showing the telltale signs of kuru—at first a drooping lower lip in Georgette, and then changes in behavior as both primates became more lethargic. Eventually the apes began to show even more clear symptoms of the disease: occasional unsteadiness and trembling followed by a gradual loss of balance.

When informed of these developments, Gajdusek was excited but cautious, worrying that the chimps might have been accidentally contaminated with scrapie. His coworkers assured him that there had been no contamination. As the researcher alternated between elation and skepticism, back in Maryland the physical deterioration of the chimps continued at a frightening pace. Only four months after the first symptoms appeared, Georgette and Daisy were almost completely paralyzed. With Gajdusek carrying out field work in one of the most isolated regions in the world, his coworkers called in a neuropathologist from London to assist with the post-mortem analysis.

On October 28, 1965, Georgette was anesthetized and sacrificed by the heartbroken researchers. Her entire body was deconstructed, fixed, and preserved, and her brain was sectioned for microscopic analysis. The results were 100 percent conclusive. Slides of Georgette's cerebellum were indistinguishable from those of human kuru victims.

Carleton Gajdusek and his colleagues had discovered a brand new disease.

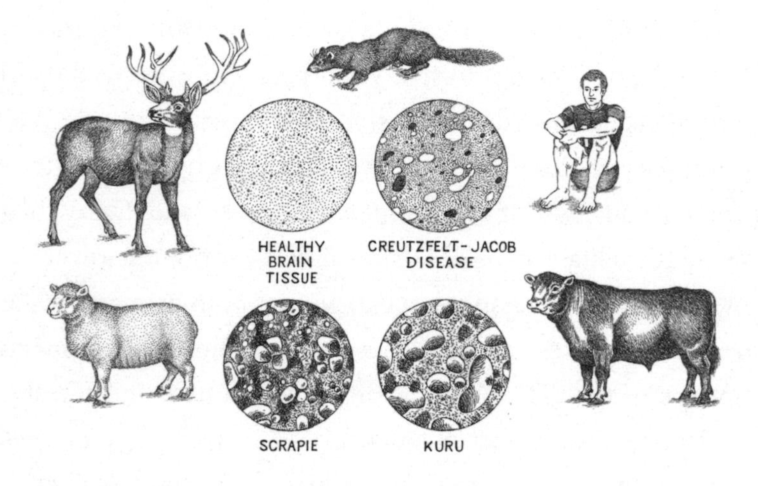


Meanwhile, Michael Alpers, who had been studying kuru since 1961 and who had taken time out from his own field work to collaborate with Gajdusek and Gibbs on the NIH primate study, waded through six years of Gajdusek's epidemiological data on the Fore. After examining hundreds of Fore genealogies, he and Gibbs came up with a remarkable observation: Instances of kuru were beginning to decline in children, starting with the youngest age group. The question immediately became, "Why?" Shortly after conferring with Robert Glasse and his wife, Shirley Lindenbaum, Alpers came up with a hypothesis.

According to information gathered from interviews with the

Fore, kuru victims were favored at mortuary feasts because the physical inactivity that characterized the latter stages of the disease left the stricken individuals with a tasty layer of subcutaneous fat. Starting in the 1950s, though, government authorities in New Guinea began cracking down on the practice of ritual cannibalism, and with mortuary feasts now forbidden by law, fewer people were eating infected tissue. As a result, incidents of the disease were decreasing. Additionally, since kuru had a shorter incubation period in children than it did in adults, the reduced occurrence of ritual cannibalism translated swiftly into a decreased incidence of kuru in the youngest sector of the population.

In a February 1966 article in *Nature*, Gajdusek, Gibbs, and Alpers described the experimental transmission of a "kuru-like syndrome" to their chimpanzees although the identity of the disease-causing agent was still unknown. Gajdusek, who still believed that they were dealing with a slow virus, was also reluctant to attribute the transmission of kuru to the consumption of infected flesh. Instead, he supported the view that during the process of handling and cutting up the dead, the kuru-causing agent was transmitted via cuts or across the thin mucous membranes that line the human mouth, eyes, and nose (a form of exposure known as inoculation).

By 1973, however, Gajdusek had come around to the idea that inoculation *and* consumption were both viable routes for kuru transmission.

> The mechanism of spread of kuru is undoubtedly contamination of the population during their ritual cannibalistic consumption of their dead relatives as a rite of respect and mourning. They did the autopsies bare-handed and did not wash thereafter; they wiped

> their hands on their bodies and in their hair, picked sores, scratched insect bites, wiped their infants' eyes, and cleaned their noses, and they ate with their hands.

Other researchers, like Joe Gibbs, stuck to the hypothesis that Fore mortuary practices, rather than the actual consumption of infected flesh, were the primary routes of kuru transmission. In a 2002 interview, the NIH researcher admitted that initial attempts to transmit kuru to chimps via a gastric tube (which modeled the consumption of infected flesh by humans) had failed, and that it was only after injecting the animals with liquefied brain material from kuru victims that they came down with the disease. As for how kuru was transmitted to the Fore, Gibbs explained that the Fore had multiple routes of inoculation, including their eyes and mouths, as well as skin lesions caused by leeches, mosquito bites, and the razor-sharp blades of *puni* grass.

Today, in regions of West Africa, the Ebola virus is often spread because of ritual practices that involve handling of recently deceased Ebola victims. For example, some Muslims believe that family members should wash the bodies of the dead, a practice that also includes the elimination of certain bodily fluids. When performed under less-than-sanitary conditions, this ritual can place individuals in grave danger if they come into contact with infectious body fluids like blood, vomit, and diarrhea—all of which characterize the advanced- and end-stage symptoms of Ebola.

I asked Shirley Lindenbaum if she thought that Fore mothers had encouraged their children to handle the dead during mortuary ceremonies.

"Mothers handed food to their small children to eat," she said.

"Since people eat with their hands, most children would touch the food given to them by their mothers and other female relatives. Children would not have been involved in the cutting of bodies, though one of my interpreters remembered sitting with others watching his mother being cut [up] and eaten. So, just as with adults, handling the food was one possible one route of infection, but as I recall, this depended on cuts and scrapes that allowed the infectious agent to enter the bloodstream—which the rest of us agree could not explain the dimensions of the epidemic. That would require a lot of cuts and scrapes, an unlikely scenario."

In October 1976, 53-year-old Daniel Carleton Gajdusek shared the Nobel Prize for Physiology or Medicine.[51] Although he was still attributing kuru to an unidentified "slow virus," other scientists had their doubts. By now, with cases of kuru dwindling to a few per year and confined to a "stone-aged society" few outsiders had ever seen, research on the disease was winding down. Interest in kuru appeared to have run its course, and with it funding for kuru-related research. Fortunately for the researchers (but unfortunately for a lot of sheep herders), scrapie, a disease that mimicked kuru's destruction of the central nervous system, was beginning to attract significant attention.

Considering the importance of the European sheep industry, it was no surprise that by the early 1970s many researchers, including Gajdusek, were pressing to understand the mechanism behind scrapie. At the forefront of the mystery was the observation that whatever the scrapie-causing agent was, it could not be killed or inactivated by disinfectants like formalin or carbolic acid.

---

51 A category usually shortened to "the Nobel Prize for Medicine."

Additionally, extracts from scrapie-infected brains lost none of their lethality after being heated, frozen, or dried. In another set of experiments, South African radiation biologist Tikvah Alper and her colleagues bombarded the mystery agent with an electron beam from a linear accelerator. Although the beam was strong enough to disrupt the molecular structure of any known pathogenic cell or virus, there was no change in the infectivity of the scrapie extract. The researchers also tried mega-doses of ultraviolet light, a proven disruptor of viral DNA and RNA—all to no avail. The extracts retained their lethality.

Alper's research team soon reached a pair of conclusions regarding the scrapie-causing agent: 1) it was far smaller than any known virus, and 2) it could replicate without nucleic acids—the chemical rungs of the helical ladder that became Watson and Crick's model for DNA. Shockingly, this last finding appeared to contradict one of the central tenets of biology, the fact that all organisms require nucleic acids to reproduce.

After reading over Alper's work, English mathematician J. S. Griffith came up with an unusual hypothesis. Perhaps, he suggested, the agent that caused scrapie wasn't a virus at all but a self-replicating protein. Griffith proposed that this mutant protein could function as a template for the production of additional mutants, each in turn taking on its own role as a template.

Researchers from competing labs scoffed at Griffith's idea and Tikvah Alper was ridiculed as a female version of virologist/biochemist Wendell Stanley, who had won a Nobel Prize in 1946 for determining that the infectious agent in Tobacco Mosaic Virus was actually a self-propagating protein—a fact that was disproven only *after* he won the award.

But Stanley Prusiner, a young biochemist out of UC San Francisco, read the papers by Alper, Griffith, and others, saw an opportunity, and jumped into the fray. In the early 1970s, Prusiner moved to Montana where his work with scrapie expert William Hadlow confirmed Alper's findings about the absence of nucleic acids in the scrapie agent. Prusiner and Hadlow's results also indicated that, when exposed to substances like enzymes that could destroy or denature proteins, the disease-transmitting ability of the scrapie extract was eliminated.

Prusiner tried to tell Gajdusek and the other NIH researchers about what he had found, but he was rebuffed. Among the kuru mavens, who were now mostly working on other projects, only Michael Alpers was supportive, inviting the American to the Goroka Institute in New Guinea, where Prusiner studied a group of nine kuru sufferers.

In 1982, Prusiner published his lab findings on scrapie in the journal *Science*. He coined the name "prion" (pronounced "PREE-on") to describe an aberrant form of protein, which he claimed was responsible for the suite of neurodegenerative disorders known as Transmissible Spongiform Encephalopathies (TSEs). Prusiner claimed that, unlike viruses, prions were not biological entities, but they could be infectious—transmitted orally or through contact with infected material. They could also be inherited or spontaneous in origin.

When asked about why the body's immune system didn't appear to mount a defense against them, Prusiner explained that unlike viruses or bacteria, prions weren't foreign invaders; they were an altered form of one of the body's own proteins. Because of this, the body never recognized them as a threat. As a result, prions would spread through the body of a TSE victim unchecked.

Prusiner did hedge his bets by stating that current knowledge did not exclude the potential existence of a small core of nucleic

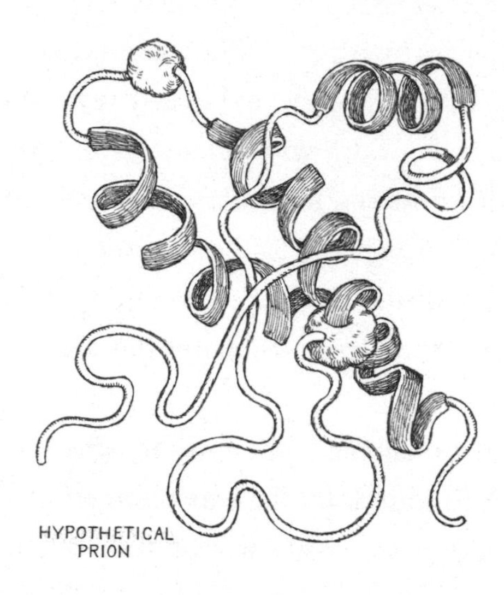


acid within the prion—which might explain how they replicated. Nevertheless, he co-opted Griffith's protein-as-template model, with his misfolded prion proteins (which were too small to see with even the most powerful microscopes) building up into the amyloid plaques that characterized fatal TSEs like kuru, Creutzfeldt-Jakob disease, TME, and scrapie.

For his work on prions, Prusiner won the Nobel Prize for Medicine in 1997. Some have argued that he should have shared the award with other researchers, and they pointed to the fact that several people had been bumped out—or worse—by the self-promoting American. In his book *Deadly Feasts*, Pulitzer Prize–winning writer Richard Rhodes wrote that Prusiner "invaded and colonized the work of others in his apparent pursuit of a Nobel Prize." Prusiner's list of enemies grew even longer after accusations that he had used the peer review process to stonewall publication of another researcher's results while submitting his own paper on a similar topic.

# 19: Acceptable Risk

*I have taken advice from the leading scientific and medical experts in this field. I have checked with them again today. They have consistently advised me in the past that there is no scientific justification for not eating British Beef and this continues to be their advice. I therefore have no hesitation in saying that beef can be eaten safely by everyone, both adults and children, including patients in hospital.*

—Sir Donald Acheson, Chief Medical Officer,
UK Department of Health, 1990

The kuru/Bovine Spongiform Encephalopathy (BSE) story now jumps to 1988.

Given the degree of bureaucracy inherent in a government such as Great Britain's, many people would have been surprised that year if the ministry had reacted to epidemiologist John Wilesmith's news on how BSE was being spread with an immediate ban on meat and bone meal, or even a warning. Instead, because they still believed that they were dealing with a disease that hadn't been transmitted to humans, the government dragged its feet. Clearly, many officials were far more concerned with preventing a panic that might impact negatively on the rendering and beef industries than they were about the possibility of their citizens consuming prion-contaminated meat pies. The government also knew that closing rendering plants would have placed the burden of eliminating unwanted livestock parts squarely on the shoulders of the beef industry, a significant new expense that would have resulted in higher meat costs and a concurrent decrease in the competitiveness of British beef on the world market. So rather than demanding immediate and industry-wide changes, the politicians did something a bit less dramatic: They quietly called for the formation of a "blue ribbon" panel led by the eminent Oxford zoologist Richard Southwood. The "Southwood Working Party" met for the first time on June 21, 1988, and again in November and December of that year. The problem was that neither Southwood nor his three-member team had any experience dealing with spongiform encephalopathies.

Earlier in June, government officials met with members of the UK Rendering Association. On the strength of the data provided by Wilesmith, the ministry informed the renderers that they would be suspending the sale of ruminant-based protein (i.e., meat and bone meal) as a dietary supplement for cows and sheep. Although the ban went into place the following month, that would become the extent of the good news. Farmers were also asked to voluntarily stop feeding meat and bone meal to their cows. Unfortunately, many of them had already spent thousands of pounds on what had suddenly become an illegal nutritional supplement. But since the government hadn't offered to buy the protein supplement back from them, and since there were no efforts to enforce the government's request, there was little incentive for the farmers to stop using it. The results were predictable.

After quietly acknowledging the fact that removing infected

cattle from the system was an important step in curtailing BSE, the ministry did decide to compensate cattle owners who turned in their visibly sick animals. But instead of offering to purchase the diseased cattle at market value, they low-balled the herd owners, offering them only 50 percent of market value for their animals. By comparison, the government was already handing out 75 percent of market value for cows infected with tuberculosis.

Ultimately, it's impossible to know just how many sick cows were hurried off to the slaughterhouse, but the numbers are thought to have been significant, especially since examples of this sort of practice are not unheard of within the meat industry, even decades later. In August 2014, Federal prosecutors indicted three Northern California slaughterhouse workers. They were charged with cutting off "USDA Condemned" stamps from sick cows and then slaughtering them while inspectors were on their lunch breaks. It is suspected that significant numbers of diseased animals were processed and sold for human consumption.

Until this time, there hadn't been much publicity about what was going on, and the British government made an effort to keep it that way. Their veil of secrecy might have remained in place far longer if several publications hadn't broken the BSE story in April 1988. The industry standard, *Farming News*, ran a front-page headline that read "Spongiform Fear Grows," while the *Sunday Telegraph* set the stage for the term "Mad Cow Disease" with a story entitled "Raging Cattle Attacks." An earlier paper in *Nature* also demonstrated that scrapie had been experimentally transmitted from sheep to monkeys, supporting Wilesmith's hypothesis that cows had gotten sick from eating scrapie-infected sheep rendered into meat and bone meal. Note: Most scientists now believe that it is more likely that

BSE originated from a spontaneous mutation in cows and did not result from sheep scrapie jumping to a new species.

The Southwood Committee published its official report in February 1989. Their most important finding supported the government's claim that they were dealing with scrapie, and so they reported that, "the risk of transmission of BSE to humans appears remote." They also concluded that there was no evidence the disease was related to Creutzfeldt-Jakob disease (CJD), the rare but deadly form of human spongiform encephalopathy.

The authors of the Southwood report also painted a rosy picture for anyone concerned about the spread of BSE, predicting that it would begin to decline in the early 1990s and die out spontaneously sometime after 1996. No further effort was required. Beef was safe. Long Live the Queen!

I MET DR. Laura Manuelidis on a beautiful summer day in New York City's Chelsea district. When she isn't writing and publishing critically acclaimed poetry, Manuelidis heads up the Neuropathology section at Yale. She has become the foremost spokesperson for the relatively few scientists who believe that prions are not the self-replicating, seemingly immortal, proteinaceous pathogens we've been led to believe. In 2007, she and her coworkers identified a 25-nanometer viruslike particle found in both sheep scrapie and CJD.

No one has ever seen a prion protein (PrP), but according to Manuelidis the reason has nothing to do with size.

"They probably don't exist," she told me.

Manuelidis has been researching neurodegenerative diseases for more than 30 years, and she and her colleagues have performed a wide range of studies on transmissible spongiform encephalopathies (TSEs) like Creutzfeldt-Jakob disease. Their results support a very different, though far from new, conclusion—that viruses are the cause of these neurodegenerative diseases, which also include kuru and bovine spongiform encephalopathy (BSE). Since the name "prion" implies infectivity (the ability of a pathogen to establish infection), in Manuelidis's book, clumps (or plaques) of misfolded proteins exist, but since they aren't infective, prions do not exist. According to Manuelidis, then, what is being called a prion requires a new name.

She explained that proteinaceous plaques aren't confined to neurological disorders like kuru, but are also seen in peripherally located viral diseases. "Conventional viruses also induce protein aggregates and amyloid fibers that are prionlike. The plaques are an end-stage product that doesn't occur early in these infections." I later learned that the abnormal protein masses were also characteristic of diseases like rheumatoid arthritis and diabetes—and in none of these instances were the clumps or the proteins that made them up transmissible.

"So what you're saying is that the misfolded proteins aren't the cause of the spongiform encephalopathy, they're an effect—a late-stage result of a viral assault. Correct?"

"Yes," she replied.

But if these plaques aren't pathogens and they're not infective, what exactly are they?

According to Manuelidis, they're "a runaway defense mechanism against the infecting agent," which she believes is viral in origin.

"When these misfolded proteins do show up, infectivity drops through the floor," she told me. In other words, once the body's defenses kick into gear (which ultimately leads to the production of amyloid plaques), the pathogen is less able to infect another host. The spread of the virus is curtailed.

On May 16, 1990, John Gummer, the head of the UK's Ministry of Agriculture, infamously responded to the public concerns over potentially contaminated beef in the UK by feeding a hamburger to his four-year-old daughter on the BBC television show *Newsnight*.

In 1993, two British dairy farmers died of CJD, a disease that was supposed to strike one out of every 1 million people. The government response was that it was just a bizarre coincidence. In May of the same year, 15-year-old British schoolgirl Victoria Rimmer began having trouble keeping her balance and within weeks she was falling constantly. Admitted to the hospital, Rimmer underwent a battery of tests—all of which came back negative. Finally a brain biopsy was obtained and the doctor who examined the results was stunned. Her brain was riddled with holes and amyloid plaques identical to those seen in the brains of kuru victims. With hesitancy, the physician informed Victoria's grandmother that the girl had spongiform encephalopathy—mad cow disease. What took place next was even more incredible. An investigator from the government's CJD-surveillance unit in Edinburgh visited Mrs. Rimmer, warning her not to say anything about her granddaughter's condition. "Think about the economy," he told her, "think about the Common Market."

Victoria Rimmer died in November 1997. After an inquest into her death, coroner John Hughes concluded that she died of natural causes.

In 1994, a 16-year-old schoolgirl and an 18-year-old boy were diagnosed with CJD, which had hardly ever been reported in people less than 30 years of age. By the following year, seven people were already dead or dying.

On March 8, 1996, the hammer fell on the government's "British Beef is Safe to Eat" stance in the form of a memo written by Dr. Eileen Rubery, a policy maker and longtime government adviser. Rubery confirmed what others had feared for 11 years—the emergence of a new form of spongiform encephalopathy, this one transmitted to humans via the consumption of contaminated beef. She also used the dreaded "e" word: epidemic. The new disease was initially referred to as sporadic CJD or atypical CJD, but scientists eventually settled on Variant Creutzfeldt-Jakob disease (vCJD).

By October 2013, the number of definite and probable deaths from vCJD in the United Kingdom stood at 177. Some researchers see the epidemic as over, pointing to the fact that after peaking in 2000, when 28 people in the UK died of vCJD, deaths from the disease have fallen off dramatically (i.e., three deaths in 2009, three in 2010, and one in 2013). Others believe that these 177 deaths are only the tip of the iceberg. They rationalize that, because thousands of Fore died as adults, sometimes 50 years after being exposed to kuru via ritual cannibalism, many Europeans (and others) who had consumed contaminated beef in the 1970s and 1980s would not have been stricken yet, and might not start dying en masse until decades after exposure.

In a 2013 study published online by the *British Medical Journal*, researchers tested 32,000 "anonymous appendix samples from people of all ages who had their appendix removed between 2000 and 2012." Sixteen of the samples, which came from 41 hospitals across England, tested positive for the abnormal prion protein. This translates into one carrier for every 2,000 people in the United Kingdom, a scary number that gets even scarier if you project that out to 493 people per 1 million inhabitants. There are approximately 63.5 million people in the United Kingdom.

On a related but more upbeat note, scientists like Simon Mead and John Collinge, both of whom are experts in the field of kuru research, think there's another reason why everyone exposed to prion-contaminated meat may not come down with a lethal neurodegenerative disease. As evidence they point to a common human gene (the prion protein gene, or PRNP) with a worldwide distribution. The researchers and their colleagues discovered a mutated form of this gene (i.e., a variant) in descendants of the Fore who survived the famous kuru outbreak. Initially, they hypothesized that this variant might have provided protection from kuru to the individuals who possessed it. These kuru-resistant survivors would have passed down their genes (and their resistance) to their descendants. In 2015, Collinge and his research team published a follow-up study in *Nature* in which they presented experimental evidence that when the genetic variant of PRNP was transferred to mice, it provided complete resistance to both kuru and classical CJD.

In a best-case scenario, thanks to what may have been their ancient ancestors' brush with cannibalism (and kuru), at least some of the individuals consuming prion-contaminated meat in the 1980s were already resistant to the disease.[52] If this is true, then the gloom-and-gloomers may be waiting for an epidemic that never arrives. From a therapeutic viewpoint, if these genetic variants can somehow be transmitted to humans, we may one day be able to confer resistance to the pathogens that cause spongiform encephalopathies—whether they turn out to be prions or viruses.

On the other hand, if Laura Manuelidis is correct and spongiform encephalopathies are the results of viruses, it would be wise to remember one of their key characteristics: Viruses mutate.

BACK IN CHELSEA, Manuelidis and I were seated in the back of a 9th Avenue bar. As she explained why viruses and not prions were the real culprits in the transmissible spongiform encephalopathy story, we sipped a pair of melon-based something-or-others. In retrospect, given the fact that my understanding of biochemistry was already more than a bit wobbly, a weak drink was probably a good thing.

"What do you make of the fact that researchers in the 1960s

---

52 According to Noel Gill, lead investigator of the "Appendix" study, further research is now underway to determine whether prion proteins also occurred in samples from the 1970s and earlier—before the appearance of BSE in the UK. Such a finding could reduce the significance of the 2013 study, since it would suggest that prion proteins in a population do not necessarily translate to a major outbreak of spongiform encephalopathy.

never saw an immune response during their experiments with kuru extracts?"

"There is an immune response," she replied. "They just didn't see it because it was transient or diminished. Like with HIV, there are many mechanisms that can make something not apparent as an immune response."

Manuelidis explained that with diseases like kuru, CJD, and BSE, "what we found is that there are innate immune responses, very early in the disease."

I shifted easily into Insistent Mode. "All right, but why couldn't prions be the TSE pathogens?"

"Because that stuff [anything composed of protein] gets digested in the gastrointestinal tract, while viruses can go through gastric and intestinal juices unscathed. Viruses are built to be protected from the body's defense mechanisms—and proteins are not."

"But what about the reports, by Prusiner and others, indicating that prions couldn't be destroyed by digestive enzymes like proteases?"

Manuelidis shook her head. She was wearing a look that suggested the question was one she'd heard plenty of times before. "Bill, everyone knows that PrP [prion protein] is relatively resistant to limited exposure to PK [a protein-dissolving enzyme found in our digestive tracts], but it breaks down if given a longer time to digest. In fact, it disappears."

I was reminded of earlier studies in which lab chimps did not become infected with kuru after being fed infected material through a gastric tube. "So you're saying that these seemingly immortal proteins *can* be broken down by our own digestive juices?"

"They're not immortal," Manuelidis replied. "And yes, all detectable forms of PrP are digested in the gastrointestinal tract, yet the invasive infectious particle [that causes TSEs like kuru] is not destroyed."

"The same way that many viruses aren't destroyed in our GI tract?"

"Right."

If this was in fact true, then how were the prions, which researchers like Prusiner claimed to be the cause of kuru, getting past a gastrointestinal tract that had evolved to digest dietary proteins? At the very least, if prions did exist, then their spread through the Fore via cannibalism no longer made sense—since their digestive tracts would have broken down the proteins during the normal process of digestion. But if the infective agent causing kuru had been a virus, then transmission through the practice of ritual cannibalism made even more sense, since viruses weren't broken down by our digestive systems.

Although the majority of the scientific community had clearly accepted prions as the pathogens behind a range of neurological diseases, there are still some strong voices besides Manuelidis's opposing this view. In a new afterword of his 1998 book, *Deadly Feasts*, Richard Rhodes admitted to being intrigued by the very real possibility that Dr. Manuelidis could be right.

"I raise the virus issue," he wrote, "partly because I now believe I gave it less than a fair hearing in the body of this book and partly because the arguments I've heard in its favor since I wrote, especially from Laura Manuelidis, seem to me compelling."

Clearly there remains much for science to explore regarding transmissible spongiform encephalopathies. One can only hope that the funding agencies that poured tens of millions of dollars into prion-related research will not see this story as a fait accompli or that researchers who challenge the very existence of self-replicating, infectious proteins won't have their voices silenced by skittish granting agencies or peer reviewers, perhaps too set in the conviction that prions are the cause of transmissible spongiform encephalopathies from kuru to mad cow disease and beyond.

# Epilogue: One Step Beyond

*Hunger hath no conscience.*

—Author unknown

Cannibalism makes perfect evolutionary sense. If a population of spiders has an abundance of males from which a female can choose, then cannibalizing a few of them may serve to increase Charlotte's overall fitness by increasing the odds that she can raise a new batch of spiderlings. On the other hand (and in spiders there are eight of these to choose from), in a population where males aren't plentiful or where the sexes cross paths infrequently, cannibalizing males would likely have a negative impact on a female's overall fitness by decreasing her mating opportunities. As a zoologist, I find this kind of dichotomy pleasing, since it's logical and appears to be more or less predictable in occurrence. In nature, as far as cannibalism is concerned, I've found no gray areas, no guilt, and no deception. There is only a fascinating variety of innocent—though often gory—responses to an almost equally variable set of environmental conditions: too many kids, not enough space, too many males, not enough food. The real complexity and the uncertainty didn't kick in until I shimmied out farther onto our own branch of the evolutionary tree. It was here that I found cannibalism painted in equal shades of red and gray.

Sigmund Freud believed that in humans, atavistic urges like cannibalism and incest are hidden below a veneer of culturally imposed taboos, and that the suppression of such forbidden behaviors signaled the birth of modern human society. This is a compelling explanation, but it's one that likely requires some serious tweaking.

Compared to other groups such as insects and fishes, cannibalism occurs less frequently in mammals and even less frequently in our closest relatives, the primates—where most examples appear to be either stress-related or due to a lack of alternative forms of nutrition. Though we humans do share some of our genetic makeup with fish, reptiles, and birds, we've evolved along a path where cultural or societal rules influence our behavior to an extent unseen in nature. Freud believed that these rules and the associated taboos prevent us from harkening back to our guilt-free and often violent animal past. Similarly, my studies have led me to conclude that the rules we've imposed in the West regarding cannibalism serve as constraints to behavior that might otherwise be deemed acceptable if we were looking at protein-starved Mormon crickets instead of indigenous Brazilians consuming their unburied dead.

There is a considerable body of evidence that cultures that were never exposed to these taboos (like *Homo antecessor*) or encountered them only relatively recently (the Chinese and the Fore of New Guinea) had no such problems undertaking a range of cannibalism-related behaviors as they developed their own sets of rules and rituals. Much to our Western dismay, some of these cultural mores extolled the virtues of cannibalism as an honor bestowed upon a deceased relative or a slain foe, or as a respectful way to treat a gravely ill parent. But even in societies where cannibalism might once have been a perfectly acceptable practice, given the pervasive influence of Western culture across the world, it's unlikely that ritual cannibalism currently exists, even on a small scale.

It's also likely that Freud would have called upon long-hidden impulses to explain our titillation with all things violent, gruesome, and forbidden. But although it's unclear to me the extent to which atavistic urges are involved, there is no doubt that we are, and seemingly have always been, fascinated by cannibalism. We need look no further than the popularity of novels like Cormac McCarthy's *The Road* (with its depiction of post-apocalyptic cannibalism), or even our obsession with vampires and zombies. A long list of popular films might begin with *The Night of the Living Dead* and its cinematic progeny, and according to *Variety*, 17.29 million viewers helped turn *The Walking Dead*'s season five premiere into the most watched cable TV show of all time.

Our language is filled with cannibal references: a woman who uses men for sex is a man-eater, while in the 1920s and 1930s a cannibal was "an older homosexual tramp who traveled with a young boy." To "eat someone" is a popular term for performing oral sex.

As for the media, consider the recent tabloid obsession with Gilberto Valle, New York City's so-called "Cannibal Cop" who was accused of conspiring to kidnap, cook, and devour his wife and another woman. "He is guilty of nothing more than having very unconventional thoughts," said Paul Gardephe, the judge who eventually overturned Valle's conviction. But besides an interesting take on "thought crimes," Valle's story also provided a glimpse into the darker corners of the Internet, where there are apparently thousands of people whose everyday fantasies revolve around cannibalizing coworkers, loved ones, or even you, Dear Reader.

Most cannibalism-related crimes, though, are thought to stem from psychological aberrations. According to forensic pathologist George Palermo, cannibal killers "are people who have a tremendous desire to destroy—a tremendous amount of hostility that they need to release. They have something stored up inside them in order to reach the point of where they want to destroy the human body and eat human flesh, and they feel a need to release that violence." Of course, such incidents are immediately condemned, although once again they often lead to fame for the cannibal and millions of dollars in revenue for those who care to recreate their stories in books or on film.

If one goes by the examples in the media ("Woman Dies After Cannibal Eats Her Face"; "Nude Face-eating Cannibal? Must Be Miami"), it would certainly seem that there are more cannibal killers out there than ever before. Even if the same percentage of cannibal killers exists now as has in the past (even the recent past), the population explosion across the planet would make it likely that there are simply more of them now. Then there's the fact that overpopulation and overcrowding are key catalysts for cannibalistic behavior in nature. Of course, some would consider it a stretch to extrapolate human behavior from the examples of spiders, fish, or hamsters. But for a zoologist, those comparisons are far less problematic.

So why the fascination with cannibalism? Or vampirism? Or serial killers? Perhaps the violent scenarios we watch and read about on a daily basis are a form of drug—one that creates excitement in lives that might otherwise be mundane and unfulfilled. According to Andrew Silke, head of criminology at the University of East

London, "Viewing anything that involves violence or death will kick-start a lot of psychological processes, such as stress and excitement. Your brain's neocortex becomes psychologically aroused, but not in a dangerous way since you're in the safe environment of your own home."

There is no definitive answer as to why cannibalism provides us with such stimulation, although what is clear, and what remains extremely disturbing for me, is our increasing desensitization to violence and gore—a trait that does not bode well for the future.

Along those lines, what would it take for cannibalism to become widespread behavior? Could it happen and, if so, how might that come about? Discounting a zombie apocalypse for now, I believe there's a scientific basis for outbreaks of widespread cannibalism, and the trigger could be something that has initiated it again and again throughout history.

The process of desertification is taking place right now in the United States, in places like Texas and even California, where researchers Daniel Griffin and Kevin Anchukaitis used soil moisture to measure drought. They determined the 2012–2014 period to be the most arid on record in 1,200 years, with 2014 coming in as the driest single year. Around the globe, across vast expanses of China, Syria, and central Africa, regions that only recently experienced dry seasons are becoming deserts. The populations of Kenya, Somalia, and Ethiopia, three of the poorest countries in the world, are suffering through the worst drought conditions in 60 years.

In the Darfur region of the Sudan (Africa's third largest country), rainfall has fallen off 30 percent over the past 40 years, and the Sahara desert is advancing into what was once farmland at a rate of one mile per year. Famine and diminishing access to fresh water

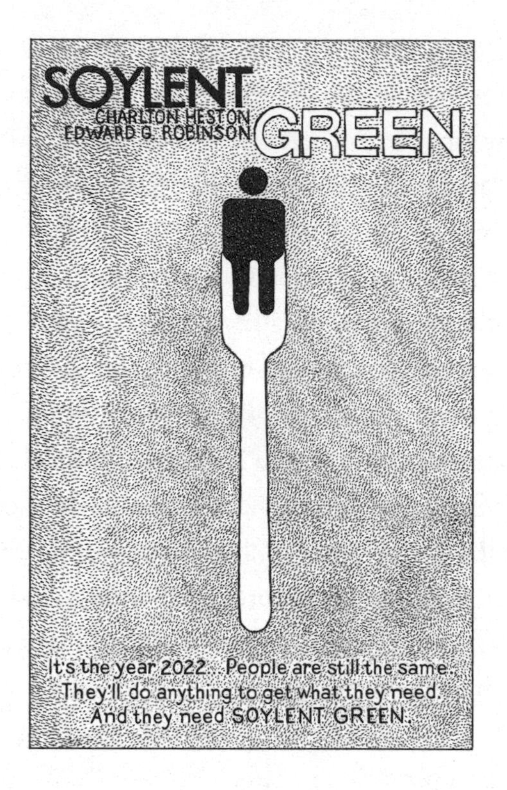


are now a daily reality for more than 12 million Africans, and these problems are growing worse every year. According to the United Nations Environmental Programme (UNEP), many of the conflicts in Africa, such as those between farmers and herdsmen, have been driven by "climate change and environmental degradation."

In 1973, Hollywood imagined just such an environmental disaster scenario in *Soylent Green*, starring Charlton Heston. His character, Frank Thorn, is a policeman in the hyper-crowded city of New York, circa 2022. Real food is now an extremely rare extravagance and most of the population subsists on nutrition wafers—including everybody's new favorite, Soylent Green. With the aid of his old friend Sol (the incomparable Edward G. Robinson, in his last role), Thorn is working on the murder of a rich Soylent Corporation executive.

During his examination of the crime scene, Thorn removes some "evidence" from the executive's apartment. This includes real food, a bottle of bourbon, and a classified oceanographic survey, dated 2016. Sol and his cronies (a group of like researchers referred to as "Books") learn that the oceans are dead and therefore unable to produce the algal protein from which Soylent Green is reputedly made. They speculate on the real ingredients and the news is not good. Heartbroken, Sol shuffles off to a government euthanasia center, downs a lethal cocktail, and dies, but not before he whispers his secret into Thorn's ear. Outside the building, the cop sneaks into the back of a truck supposedly transporting the bodies of the euthanized to a crematorium, but instead it heads straight to the Soylent manufacturing facility where Sol's dying words are confirmed.

"They're making our food out of people!" Thorn tells a fellow cop (after the requisite gun battle). "Next thing they'll be breeding us like cattle." Seriously wounded, Thorn is carried away on a stretcher, screaming what would become the American Film Institute's 77th most famous quote in movie history.

"Soylent Green is people!"

Though the special effects are dated and the action is reduced to the standard cop chases the bad guys, *Soylent Green* remains a scary 1970s take on an Earth ravaged by climate change, pollution, and overpopulation. It is a nightmare vision that comes complete with government-sanctioned cannibalism—embraced and efficiently carried out by Big Business.

And now for a bit of 21st-century speculation, which might

sound hypocritical coming from someone who's been trying to avoid cannibal-related sensationalism for the past 19 chapters.

Since cannibalism is a completely normal response to severe stress, especially during times of famine and warfare, how much of a surprise would it be if the butchery of humans for food becomes commonplace in drought-ridden and overpopulated regions of the near-future Earth? According to sociologist Pitirim Sorokin, famine-related cannibalism occurred 11 times in Europe between 793 CE and 1317 CE, as well as in "ancient Egypt, ancient Greece and Rome, Persia, India, China and Japan."

In a world where global climate change is taking place before our very eyes, there may be little to prevent famine-related cannibalism from happening again, especially in the poorest and most unstable countries in the world. Even scarier, knowing what we know about spongiform encephalopathies like kuru, as well as what we don't know, the consequences of widespread cannibalism would likely extend beyond the horrific images that would find their way onto the nightly news and social media. Could something like kuru become an epidemic on an even wider scale than it did in New Guinea? According to biological anthropologist Simon Underdown, it might have happened already.

Underdown thinks that it is possible that not all of the victims of transmissible spongiform encephalopathies (TSEs) were burger-chomping Brits or New Guineans whose Stone Age lifestyles lasted well into the 20th century. He suggests that some of those victims may actually have lived in the Stone Age, and that their disappearance some 30,000 years ago may have been hastened by the widespread practice of cannibalism and a kuru-like epidemic that resulted from the behavior.

Underdown summarized his hypothesis by first reminding me that "most of us have shifted to the idea that there were a number of Neanderthal extinctions, taking place at different times and at different locations." There is also evidence, he said, that Neanderthals "engaged in cannibalism to some degree." Having established a potential route for the contraction of transmissible spongiform encephalopathy, Underdown suggested that, in addition to the actual consumption of diseased individuals, disease transmission might have occurred through the use of stone tools contaminated by the blood and tissue of infected individuals. Like Shirley Lindenbaum's explanation for how kuru was spread along trade routes and through the movement of individuals between groups, Underdown hypothesized that similar actions could have introduced "new clusters of TSE infection" into the small, spread-out groups of Neanderthals.

Underdown then described a model he designed to test the effect of a disease like kuru on Neanderthal population numbers, which he estimated to be in the range of 10,000 to 20,000 individuals at any one time. The results of his "Kuru Model" suggested that deaths from a kurulike TSE "could reduce the population to non-viable levels within the space of 250 years."

To be clear, Underdown wasn't claiming that kuru was the sole reason that the Neanderthals disappeared. If one imagined, though, that a kurulike disease arose on multiple occasions within small populations, the effect would have been more localized, "a drip, drip effect" rather than a massive extinction event. Perhaps the invasion of the Neanderthal homeland by modern humans served as the coup de grâce to a species whose cannibalistic habits had already brought them to the brink of extinction.

Of course, any future outbreaks of kuru would occur in a world that has seen tremendous advances in many fields of medicine. But on the other side of the ledger, one need only look at the massive death toll from recent Ebola epidemics to realize that cannibalism-related outbreaks of kuru could have a devastating effect on local populations. The debate over what causes kuru and other transmissible spongiform encephalopathies is still ongoing within the scientific community. Are TSEs caused by a virus or by prions? The argument would certainly go public if desertification and global climate change (or perhaps another environmental disaster) led to outbreaks of cannibalism and the associated neurological diseases. With two Nobel Prizes already awarded for research into TSEs, in all likelihood someone would win another for finally figuring it all out. And what would they call this new cannibalism-related nightmare? Mad cow disease and the laughing death are already taken. The new strain, with its potential for killing on a scale unprecedented for transmissible spongiform encephalopathies, would need its own name—something lurid. And the only certainty is that someone will come up with one.

# Acknowledgments

I'd like to thank my agent Gillian MacKenzie for her hard work, great advice, and perseverance in getting my book off the ground. Thanks also to Kirsten Wolff and Allison Devereux of the Gillian MacKenzie Agency for their assistance, especially during the bumpy stretches.

I also offer my sincere thanks to Amy Gash, my incredibly talented editor at Algonquin. Thanks also go out to the entire production and marketing teams at Algonquin Books, especially my copy editor, Marina Lowry.

I was very lucky to have worked with or received assistance from a long list of experts who were extremely generous with their time. Thanks and gratitude go out to: Cristo Adonis, Stephen Amstrup, William Arens, Ronald Chase, Ken Dunn, Rainer Foelix, Laurel Fox, John Grebenkemper (and his dog Kayle), Kristin Johnson (Donner Party expert extraordinaire), Mary Knight, Walter Koenig, Mark Kristal, Nick Levis, Shirley Lindenbaum, John Lurie, Laura Manuelidis, Ryan Martin, Mark Norell, David and Karin Pfennig, Clair and William Rembis (and the 11 "Rembi"), Raymond Rogers,

Antonio Serrato, Stephen Simpson, Ian Tattersall, Simon Underdown, Marvalee Wake, Jerome Whitfield, and Mark Wilkinson.

I owe a huge debt of gratitude to my friends and colleagues in the bat research community and at my favorite place in the world, the American Museum of Natural History. They include Ricky Adams, Wieslaw Bogdanowicz, Frank Bonaccorso, Mark Brigham, Patricia Brunauer (RIP), Deanna Byrnes, Catherine Doyle-Capitman, Betsy Dumont, Neil Duncan, Nicole Edmison, Arthur Greenhall (RIP), "Uncle" Roy Horst, Tigga Kingston, Mary Knight (who told me to blame it on the Greeks), Karl Koopman (RIP), Tom Kunz, Gary Kwiecinski, Ross MacPhee, Eva Meade and Rob Mies (Organization for Bat Conservation), Shahroukh Mistry, Maceo Mitchell, Mike Novacek, Ruth O'Leary, Stuart Parsons, Scott Pedersen, Nancy Simmons (it's good to know the Queen), Elizabeth Sweeny, Ian Tattersall, Merlin Tuttle, Rob Voss, John Walhert, and Eileen Westwig.

I've been fortunate to have had several incredible mentors, none more important than John W. Hermanson (Cornell University, Field of Zoology). John took a chance on me in 1990 by taking me on as his first Ph.D. student. Among other things, he taught me to think like a scientist, as well as the value of figuring things out for myself. "Here's to you, Chief."

A very special thanks to my great friend, confidant, and co-conspirator, Leslie Nesbitt Sittlow, with whom I spent many hours discussing the pros and cons of cannibalism (among other things).

My dear friends Darrin Lunde and Patricia J. Wynne were instrumental in helping me develop this project from a vague idea into a finished work. A millions thanks also to Patricia for all of the amazing figures. I can't wait for our next project.

A special thank you goes out to my teachers, readers, and

supporters at the Southampton College Summer Writers Conference, especially Bob Reeves, Bharati Mukherjee, and Clark Blaise.

At Southampton College (RIP) and LIU Post, thanks and gratitude to Ted Brummel, Scott Carlin, Matt Draud, Gina Famulare, Paul Forestell, Art Goldberg, Katherine Hill-Miller, Jeff Kane, Kathy Mendola, Howard Reisman, Jen Snekser, and Steve Tettlebach. Thanks also to my LIU graduate students: Maria Armour, Aja Marcato, Megan Mladinich, and Chelsea Miller.

Sincere thanks goes out to the Adamo family, John E.A. Bertram, John Bodnar, Chris Chapin, Alice Cooper, Suzanne Finnamore Luckenbach (who predicted it all), John Glusman, Kim Grant, The Peconic Land Trust, Gary Johnson, Kathy Kennedy, Bob Lorzing, my former agent—the legendary and wonderful Elaine Markson, Carrie McKenna, Farouk Muradali (my friend and mentor in Trinidad), Erin Nicosia (Amercian Cheese), the Pedersen family and various offshoots, Gerard, Oda and Dominique Ramsawak (for their friendship and all things Trinidadian), Isabella Rossellini, Jerry Ruotolo (my great friend and favorite photographer), Laura Schlecker, Richard Sinclair, Edwin J. Spicka (my mentor at the State University of New York at Geneseo), and Katherine Turman (*Nights with Alice Cooper*). Special thanks also go out to Mrs. Dorothy Wachter, who listened patiently and provided encouragement, nearly forty years ago, when I told her I wanted to write books. I think she would have had a great laugh at the topic of this one.

Finally, my eternal thanks and love go out to my family for their patience, love, encouragement, and unwavering support, especially Janet Schutt, Billy Schutt, Chuck and Eileen Schutt, Bobby and Dee Schutt, Rob, Shannon and Kelly Schutt, my cousins, nieces and nephews, my grandparents (Angelo and Millie DiDonato), all my Aunt Roses, and of course, my late parents, Bill and Marie Schutt.

# Notes

## Prologue

ix **To mark its 100-year anniversary in 2003:** This information appears on AFI's website, http://www.afi.com/100years/handv.aspx.

x **Stefano's screenplay for *Psycho*:** Robert Bloch, *Psycho* (New York: Random House, 1958).

xi **"She isn't missing," Gein told them:** Francis Miller and Francis Scherschel (photographs), "House of horror stuns the nation," *Life* 43, no. 23 (December 2, 1957): 24–32, p. 28.

xi **and a three-page spread in *Time*:** *Time*, "Portrait of a killer," 70, no. 23 (December 2, 1957): 32–33.

xii **"I had a compulsion to do it":** Moira Martingale, *Cannibal Killers* (New York: Carroll & Graf Publishers, 1993), p. 81.

xv **"Polar bears resort to cannibalism:** Marsh Walton, "Polar bears resort to cannibalism as Arctic ice shrinks," *CNN Tech*, September 23, 2008, http://articles.cnn.com/2008-09-23/tech/arctic.ice_1_sea-ice-national-snow-ice-data-center?_s=PM:TECH.

xv **"Climate Change Forcing Polar Bears:** Foreign Staff, "Climate change 'forcing polar bears to become cannibals,'" *Times* (London), December 9, 2009, http://www.timesonline.co.uk/tol/news/environment/article6949625.ece.

xv **It was Reuters, though:** I. Williams, "Polar bear turns cannibal," Reuters, December 9, 2009, http://www.reuters.com/news/pictures/slideshow?articleId=USRTXRLWU#a=1.

## Chapter 1—Animal the Cannibal

5 **"cannibalism was also a consistent part:** Lauren Fox, personal communication.

5 **By the time Fox's review paper:** L. Fox, "Cannibalism in natural populations," *Annual Review of Ecology and Systematics* 6 (1975): 87–106.

5 **many that were long considered to be herbivores:** Ibid., p. 88.

6 **In 1980, ecologist and scorpion expert:** G. Polis, "The evolution and dynamics of intraspecific predation," *Annual Review of Ecology and Systematics* 12 (1981): 225–51.

6 **Fox even followed cannibalism's environmental connection:** Fox, "Cannibalism in natural populations," p. 89.

7 **In 1992, zoologists Mark Elgar and Bernard Crespi:** M. A. Elgar and B. J. Crespi, eds., *Cannibalism: Ecology and Evolution among Diverse Taxa* (New York: Oxford University Press, 1992).

7 **In it, they refined the scientific definition:** Ibid., p. 2.

12 **David Pfennig and his colleagues proposed:** D. W. Pfennig, "The adaptive significance of an environmentally-cued developmental switch in an anuran tadpole," *Oecologia* 85 (1990): 101–7.

12 **Pfennig hypothesized that iodine-containing:** D. W. Pfennig, "Proximate and functional causes of polyphenism in an anuran tadpole," *Functional Ecology* 6 (1992): 167–74.

13 **Pfennig and his coworkers previously:** D. W. Pfennig, P. W. Sherman, and J. P. Collins, "Kin recognition and cannibalism in polyphenic salamanders," *Behavioral Ecology* 5 (1994): 225–32.

15 **In species like the flour beetle:** F. K. Ho and P. Dawson, "Egg cannibalism by *Tribolium* larvae," *Ecology* 47 (1966): 318–22.

## Chapter 2—Go On, Eat the Kids

21 **These prepackaged meals:** T. M. Spight, "On a snail's chances of becoming a year old," *Oikos* 26 (1975): 9–14.

22 **The black lace-weaver spider:** K. W. Kim and C. Roland, "Trophic egg laying in the spider, *Amaurobius ferox*; mother-offspring interactions and functional value," *Behavioral Processes* 50, no. 1 (2000): 31–42.

22 **The ravenous larva of the elephant mosquito:** P. S. Corbett and A. Griffiths, "Observations of the aquatic stages of two species of *Toxorhynchites* (Diptera: Culicidae) in Uganda," *Proceedings of the Royal Entomological Society of London A* 38 (1963): 125–35.

23 **In some snail species:** B. Baur, "Effects of early feeding experience and age on the cannibalistic propensity of the land snail *Arianta arbustorum*," *Canadian Journal of Zoology* 65 (1987): 3068–70.

24 **In mammals, filial cannibalism:** R. Elwood, "Pup-cannibalism in rodents: Causes and consequences," in *Cannibalism: Ecology and Evolution among Diverse Taxa*, edited by Mark Elgar and Bernard Crespi (New York: Oxford University Press, 1992), pp. 299–322.

24 **In the fishes, by far the largest:** G. FitzGerald and F. Whoriskey, "Empirical studies of cannibalism in fish," in Elgar and Crespi, *Cannibalism*, p. 251.

26 **But even in the 90 or so:** Ibid., pp. 244–45.

29 **But in sand tiger sharks:** R. G. Gilmore, J. W. Dodrill, and P. A. Linley, "Reproduction and embryonic development of the sand tiger shark, *Odontapsis taurus* (Rafinesque)," *Fishery Bulletin* 8, no. 2 (1983): 201–25.

29 **These embryos (averaging 19:** G. Polis, "The evolution and dynamics of intraspecific predation," *Annual Review of Ecology and Systematics* 12 (1981): 241.

31 **Cannibalism of the young also occurs:** G. Polis and C. Myers, "A survey of intraspecific predation among reptiles and amphibians," *Journal of Herpetology* 19, no. 1 (1985): 99.

31 **significant juvenile mortality in the American alligator:** M. Delany, A. Woodward, R. Kiltie, and C. Moore, "Mortality of American alligators attributed to cannibalism," *Herpetologica* 67, no. 2 (2011): 174–85.

31 **Cannibalism is relatively rare among birds:** D. Mock, "Infanticide, siblicide, and avian nestling mortality," in *Infanticide*, edited by Glenn Hausfeter and Sarah Blaffer Hrdy (Hawthorne, NY: Aldine Publishing, 1984), p. 6.

32 **In one study of a colony of 900 herring gulls:** J. Parsons, "Cannibalism in herring gulls," *British Birds* 64 (1971): 528–37.

32 **Sibling cannibalism, in which brothers:** M. Stanback and W. Koenig, "Pup cannibalism in rodents," in *Cannibalism: Ecology and Evolution among Diverse Taxa*, edited by Mark Elgar and Bernard Crespi (New York: Oxford University Press, 1992), pp. 285–86.

33 **filial cannibalism in birds:** Ibid., pp. 285–87.

## Chapter 3—Sexual Cannibalism, or Size Matters

36 **Back then, several authors claimed:** Two examples are L. O. Howard, "The excessive voracity of the female mantis," *Science* 8 (1886): 326; J. H. Fabré, *Souvenirs Entomologiques*, vol. 5 (Paris: De Lagrave, 1897).

37 **One hypothesis reasoned that the male mantis's brain:** K. D. Roeder, "An experimental analysis of the sexual behavior of the praying mantis (*Mantis religiosa*)," *Biological Bulletin* 69 (1935): 203–20.

39 **They believe that, rather than acting as a stimulus:** E. Liske and W. J. Davis, "Courtship and mating behavior of the Chinese praying mantis, *Tenodera sinensis*," *Animal Behavior* 35 (1987): 1524–37.

39 **After several papers in the 1930s and 1940s:** Two examples are W. B. Herms, S. F. Baily, and B. McIvor, "The black widow spider," *California Agricultural Experimental Station Bulletin* 59 (1935): 1–29; R. N. Smithers, "Contribution to our knowledge of the genus *Latrodectus* in South Africa," *Annals of the South African Museum* 36 (1944): 263–312.

40 **They determined that not only did most male spiders:** R. F. Foelix, *Biology of Spiders* (Cambridge, MA: Harvard University Press, 1982), p. 316.

40 **sexual cannibalism has been reported in 16:** M. Elgar, "Sexual cannibalism in spiders and other invertebrates," in *Cannibalism: Ecology and Evolution among Diverse Taxa*, edited by Mark Elgar and Bernard Crespi (New York: Oxford University Press, 1992), pp. 129–43.

41 **The next phase of redback courtship begins:** L. M. Forster, "The stereotypical behavior of sexual cannibalism in *Latrodectus hasselti* (Araneae: Theridiidae), the Australian redback spider," *Australian Journal of Zoology* 40 (1992): 1–11.

44 **Mark Elgar and zoologist David Nash worked with this species:** M. Elgar and D. Nash, "Sexual cannibalism in the garden spider, *Araneus diadematus*," *Animal Behavior* 36 (1988): 1511–71.

45 **To determine why, arachnologist Anita Aisenberg:** A. Aisenberg, F. Costa, and M. Gonzalez, "Male sexual cannibalism in a sand-dwelling wolf spider with sex role reversal," *Biological Journal of the Linnean Society* 103, no. 1 (2011): 68–75.

46 **Cannibalism by males also occurs in the aptly named water spiders:** D. Schütz and M. Taborsky, "Mate choice and sexual conflicts in the size

dimorphic water spider *Argyroneta aquatica* (Araneae, Argyronetidae)," *Journal of Arachnology* 33 (2005): 767–75.

48 **banana slugs (genus *Ariolimax*):** H. Reise and J. Hutchinson, "Penis-biting slugs: Wild claims and confusions," *Trends in Ecology and Evolution* 17, no. 4 (2002): 163.

50 **Chase and his colleagues showed that this hormonelike substance:** R. Chase and K. Blanchard, "The snail's love-dart delivers mucus to increase paternity," *Proceedings of the Royal Society B* 273 (2006): 1471–75.

50 **Additionally, a 2013 study by Japanese researchers:** K. Kimura, K. Shibuya, and S. Chiba, "The mucus of a land snail love-dart suppresses subsequent matings in darted individuals," *Animal Behavior* 85 (2013): 631–35.

## Chapter 4—Quit Crowding Me

53 **According to biologist and Mormon cricket expert:** S. J. Simpson, G. Sword, P. Lorch, and I. Couzin, "Cannibal crickets on a forced march for protein and salt," *Proceedings of the National Academy of Sciences* 103, no. 11 (2006): 4152–56.

53 **Simpson and his coworkers conducted food preference tests:** Ibid.

54 **Initially, cannibalism on poultry farms:** R. Trudelle-Schwarz, "Cannibalism: Chicken Little meets Hannibal Lector?" *Behave: Stories of Applied Animal Behavior*, http://www.webpages.uidaho.edu/range556/appl_behave/projects/chicken_cannibalism.html (accessed September 12, 2016).

55 **researchers now believe that it's actually misdirected foraging behavior:** B. Huber-Eicher and B. Wechsler, "The effect of quality and availability of foraging materials on feather pecking in laying hen chicks," *Animal Behaviour* 55 (1998): 861–73.

56 **The results of a study on golden hamsters:** R. Gattermann, R. Johnston, N. Yigit, P. Fritzsche, S. Larimer, S. Özkurt, K. Neumann, Z. Song, E. Colak, J. Johnson, and M. McPhee, "Golden hamsters are nocturnal in captivity but diurnal in nature," *Biology Letters* 4 (2008): 253–55.

57 **As a result of this laundry list:** R. Elwood, "Pup-cannibalism in rodents: Causes and consequences," in *Cannibalism: Ecology and Evolution among Diverse Taxa*, edited by Mark Elgar and Bernard Crespi (New York: Oxford University Press, 1992), pp. 299–322.

58 ***M. auratus* has the shortest gestation period:** Encyclopedia of Life, http://eol.org/pages/1179513/details.

58 **When non-human primates (i.e., monkeys and apes):** M. Hiraiwa-Hasegawa, "Cannibalism among non-human primates," in Elgar and Crespi, *Cannibalism*, pp. 323–38.

59 **"A female who loses her infant:** J. Goodall, "Infant killing and cannibalism in free-living chimpanzees," *Folia primatologica* 28 (1977): 271.

59 **Other attacks by male chimps on infant-baring females:** Ibid., p. 260.

60 **Goodall believes that the attacks:** Ibid., p. 269.

61 **A team led by comparative psychologist:** S. Anitei, "Female chimps practice heavily infanticide and cannibalism," *Softpedia*, May 15, 2007, http://news.softpedia.com/news/Female-Chimps-Practice-Heavily-Infanticide-and-Cannibalism-54687.shtml.

## Chapter 5—Bear Down

62 **"Polar Bears Are Turning to Cannibalism:** B. Johnson, "Polar bears are turning to cannibalism as Arctic ice disappears, " *Think Progress*, December 8, 2011, http://thinkprogress.org/climate/2011/12/08/385037/polar-bears-are-turning-to-cannibalism-as-arctic-ice-disappears/.

62 **"Is Global Warming Driving Polar Bears:** N. Wolchover, "Is global warming driving polar bears to cannibalism?" *Live Science*, December 15, 2011, http://www.livescience.com/17500-global-warming-driving-polar-bears-cannibalism.html.

62 **"Polar Bear Cannibalism Linked to:** O. Grigoras, "Polar bear cannibalism linked to climate change," *Softpedia*, December 9, 2011, http://news.softpedia.com/news/Polar-Bear-Cannibalism-Linked-to-Climate-Change-239585.shtml.

63 **Cannibalism has been recorded in at least 14:** G. Polis, "The evolution and dynamics of intraspecific predation," *Annual Review of Ecology and Systematics* 12 (1981): 231.

63 **Heterocannibalism, in this case, eating the cubs:** B. Bertram, "Social factors influencing reproduction in wild lions," *Journal of Zoology* 177 (1975): 463–82.

66 **"Polar bears will readily eat other polar bears:** M. Taylor, T. Larsen, and R. Schweinsburg, "Observations of intraspecific aggression and

cannibalism in polar bears (*Ursus maritimus*)," *Arctic* 38, no. 4 (1985): 303.

67 **The mess came about soon after the 2006:** S. C. Amstrup, I. Stirling, T. Smith, C. Perham, and G. Thiemann, "Recent observations of intraspecific predation and cannibalism among polar bears in the southern Beaufort Sea," *Polar Biology* 29 (2006): 997–1002.

67 **"The underlying causes for our cannibalism:** Ibid., p. 1001.

67 **"chance observations of previously unobserved:** Ibid.

67 **"the first population segment to show:** Ibid.

68 **the first published report surfacing in 1897:** F. Nansen, *Farthest North*, vol. 2 (London: MacMillan and Company, 1897), 254–56.

68 **"GRAPHIC PHOTOS":** J. Zelman, "Polar bear eats cub: Cannibalism may be on the rise (GRAPHIC PHOTOS)," *Huffington Post*, December 8, 2011, http://www.huffingtonpost.com/2011/12/08/polar-bear-eats-cub-cannibalism_n_1136428.html.

## Chapter 6—Dinosaur Cannibals?

69 **In 1947, a team from the American Museum of Natural History:** S. Nesbitt, A. Turner, G. Erickson, and M. Norell, "Prey choice and cannibalistic behavior in the theropod *Coelophysis*," *Biological Letters* 2 (2006): 611.

70 **Led by paleontologists Sterling Nesbitt:** Ibid., pp. 611–14.

71 **In a much-publicized case:** Aase Jacobsen, "Feeding behavior of carnivorous dinosaurs as determined by tooth marks on dinosaur bones," *Historical Biology* 13, no. 1 (1998): 17–26.

71 **They concluded that while cannibalism:** Nesbitt et al., "Prey choice and cannibalistic behavior in the theropod *Coelophysis*," p. 614.

71 **"cannibalism seems to have been a surprisingly common:** N. Longrich, J. John Horner, G. Erickson, and P. Currie, "Cannibalism in *Tyrannosaurus rex*," *PLoS ONE* 5, no. 10 (2010): e13419, http://www.plosone.org/article/info%3Adoi%2F10.1371%2Fjournal.pone.0013419.

73 **the only compelling evidence for dinosaur cannibalism:** R. Rogers, D. Krause, and K. Curry Rogers, "Cannibalism in the Madagascan dinosaur, *Majungatholus atropus*," *Nature*, 422 (2003): 515–18.

## Chapter 7—File Under: Weird

79 **All caecilians do share one characteristic:** M. Wake, "Fetal maintenance and its evolutionary significance in the Amphibia: Gymnophiona," *Journal of Herpetology* 11, no. 4 (1977): 384.

80 **she referred to as "secretory beds":** Ibid., p. 380

80 **Parker had previously labeled "uterine milk":** Ibid.

80 **"a thick white creamy material:** Ibid.

81 **Wake proposed that fetal caecilians:** Ibid.

81 **In 2006, caecilian experts:** A. Kupfer, H. Müller, M. Antoniazzi, C. Jared, H. Greven, R. Nussbaum, and M. Wilkinson, "Parental investment by skin feeding in a caecilian amphibian," *Nature* 440 (2006): 926–29.

82 **"The outer layer is what they eat":** J. Owen, "Flesh-eating baby 'worm' feasts on mom's skin," *National Geographic News*, April 12, 2006.

83 **It also explained why mothers:** Kupfer et al., "Parental investment by skin feeding in a caecilian amphibian," p. 927.

83 **Scientists now believe:** Ibid., p. 928.

## Chapter 8—Neanderthals and the Guys in the Other Valley

85 **These anatomical differences led him to conclude:** R. Roger Lewin and R. Foley, *Principles of Human Evolution* (Oxford, UK: Wiley-Blackwell, 2004), p. 395.

86 **Huxley announced that *Homo sapiens* had descended:** T. H. Huxley, *Evidence as to Man's Place in Nature* (London: Williams & Norgate, 1864), p. 159.

87 **The anthropologist also claimed:** Ian Tattersall, *The Fossil Trail* (New York: Oxford University Press, 2009), p. 47.

87 **In "The Grisly Folk":** H. G. Wells, "The grisly folk," *Storyteller Magazine*, April 1921, http://www.trussel.com/prehist/grisly.htm.

88 **"Every feature that Boule stressed:** N. Eldridge and I. Tattersall, *The Myths of Human Evolution* (New York: Columbia University Press, 1982), p. 76.

88 **what is known as the Regional Continuity hypothesis:** M. H. Wolpoff, J. Hawks, R. Caspari, "Multiregional, not multiple origins," *American Journal of Physical Anthropology* 112 (2000): 129–36.

90 **Supporting this stance are recent morphological**: E. Kranioti, R. Holloway, S. Senck, T. Ciprut, D. Grigorescu, and K. Harvati, "Virtual assessment of the endocranial morphology of the early modern european fossil Calvaria from Cioclovina, Romania," *Anatomical Record* 294, no. 7 (2011): 1083–92.

90 **mitochondrial DNA studies that indicate:** J. P. Noonan, "Neanderthal genomics and the evolution of modern humans," *Genome Research* 20, no. 5 (2010): 547–53.

90 **Further support came from:** M. Currat and L. Excoffier, "Strong reproductive isolation between humans and Neanderthals inferred from observed patterns of introgression," *Proceedings of the National Academy of Science* 108, no. 37 (2011): 15129–34.

91 **briefly argued in 1866:** Ian Tattersall, *The Last Neanderthal* (New York: Macmillan, 1996), p. 88.

91 **researchers now believe that a hyena:** T. White, N. Toth, P. Chase, G. Clark, N. Conrad, J. Cook, F. d'Errico, R. Donahue, R. Gargett, G. Giacobini, A. Pike-Tay, and A. Turner, "The question of ritual cannibalism at Grotta Guattari [and comments and replies]," *Current Anthropology* 32, no. 2 (1991): 118–38.

91 **multiple sites in northern Spain, southeastern France, and Croatia:** Ibid., p. 58-65.

92 **"Bodies may be buried, burned, placed:** T. White, "Once were cannibals," *Scientific American*, August 2001, p. 61.

93 **Archaeologists now consider this type:** Ibid., p. 62.

93 **An excavation begun there in 1991:** A. Defleur, T. White, P. Valensi, L. Slimak, and E. Crégut-Bonnoure, "Neanderthal cannibalism at Moula-Guercy, Ardèche, France," *Science* 286 (1999): 128–31.

94 **In 2000, researchers working:** R. Marlar, B. Leonard, B. Billman, P. Lambert, and J. Marlar, "Biochemical evidence of cannibalism at a prehistoric Puebloan site in southwestern Colorado," *Nature* 401 (2000): 74–78.

94 **It is a finding that has been:** K. Dongoske, D. Martin, and T. Ferguson, "Critique of the claim of cannibalism at Cowboy Wash," *American Antiquity* 65, no. 1 (2000): 179–90.

95 **One instance in which the evidence:** Y. Fernández-Jalvo, J. Diez, I. Cáceres, J. Rosell, "Human cannibalism in the early Pleistocene of Europe (Gran Dolina, Sierra de Atapuerca, Burgos, Spain)," *Journal of Human Evolution* 37 (1999): 591–622.

96 **The first fossils of this species:** Tattersall, *The Fossil Trail*, p. 228.

96 **Excavation of the pit:** Ibid., p. 229.

96 **the site has yielded more than 5,000 bone fragments:** Ibid., p. 229.

96 **including a large pelvis:** Ibid., p. 230.

96 **"tool-induced surface modification":** Fernández-Jalvo et al., "Human cannibalism in the early Pleistocene of Europe," p. 599.

97 **"the victims of other humans:** Ibid., p. 620.

## Chapter 9—Columbus, Caribs, and Cannibalism

99 **The captain . . . took two parrots:** P. Hulme, "Introduction: The cannibal scene," in *Cannibalism and the Colonial World*, edited by Francis Barker, Peter Hulme, and Margaret Iversen (Cambridge, UK: Cambridge University Press, 1998), p. 16.

100 **"[The Arawaks] are fitted to be ruled:** U. Bitterli and R. Robertson, *Cultures in Conflict: Encounters Between European and Non-European Cultures, 1492–1800* (Stanford, CA: Stanford University Press, 1993), p. 75.

101 **these warrior women lived on their own island:** C. Sauer, *The Early Spanish Main* (Berkeley, CA: University of California Press, 1966), p. 23.

101 **preparing their viands by smoking them:** R. Tannahill, *Flesh & Blood: A History of the Cannibal Complex* (London: Little, Brown, 1975, 1996), p. 108.

102 **some Caribs had doglike faces:** F. Lestringant, *Cannibals: The Discovery and Representation of the Cannibal from Columbus to Jules Verne*, translated by Rosemarie Morris (Berkeley, CA: University of California Press, 1997), pp. 15–19.

103 **the transition from Carib to Canib:** D. Korn, M. Radice, and C. Hawes, *Cannibal: The History of the People-Eaters* (London: Channel 4 Books, 2001), p. 11.

103 **I agree with Yale professor:** C. Rawson, "Unspeakable rites: Cultural reticence and the cannibal question," *Social Research* 66, no. 1 (1999): 167–93.

105 **"why God Our Lord:** Sauer, *The Early Spanish Main*, p. 98.

105 **If such cannibals continue to resist:** N. Whitehead, "Carib cannibalism: The historical evidence," *Journal de la Société des Américanistes* 70 (1984): 69–88.

105 **Pope Innocent IV decreed in 1510:** Ibid., p. 72.

105 **On islands where no cannibalism had been reported:** W. Arens, *The Man-Eating Myth* (Oxford: Oxford University Press, 1979), p. 51.

106 **Rodrigo de Figueroa, the former governor:** Whitehead, "Carib cannibalism," p. 71.

107 **According to historian David Stannard, "Wherever the marauding:** D. E. Stannard, *American Holocaust: Conquest of the New World* (Oxford: Oxford University Press, 1992), p. 69.

107 **The diseases the Spaniards carried:** Ibid., p. 68.

107 **Stannard believes that by the end:** Ibid., p. 95.

107 **Political scientist Rudolf Rummel estimates:** R. J. Rummel, *Death by Government* (New Brunswick, NJ: Transaction Publishers, 1994), chap. 3, http://www.hawaii.edu/powerkills/DBG.CHAP3.HTM.

## Chapter 10—Bones of Contention

110 **anthropologist Neil Whitehead suggests:** N. Whitehead, "Carib cannibalism: The historical evidence," *Journal de la Société des Américanistes* 70 (1984): 69–88.

111 **"The ordinary food of the Caribs:** Ibid., p. 77.

111 **ritualized cannibalism can be differentiated**: L. R. Goldman, "From pot to polemic: Uses and abuses of cannibalism," in *The Anthropology of Cannibalism*, edited by Laurence R. Goldman (Westport, CT: Bergin and Garvey, 1999), p. 44.

112 **In the Pacific Theater during World War II:** C. Hearn, *Sorties Into Hell: The Hidden War on Chichi Jima* (Santa Barbara, CA: Greenwood Publishing Group, 2003), p. 226.

113 **The lucky man's name was Lt. George H. W. Bush:** C. Laurence, "George Bush's comrades eaten by their Japanese POW guards," *Telegraph*, October 26, 2003.

113 **Anthropologist Beth Conklin studied the Wari':** B. A. Conklin, *Consuming Grief: Compassionate Cannibalism in an Amazonian Society* (Austin, TX: University of Texas Press, 2001), p. 368.

113 **"Wari' are keenly aware that prolonged grieving:** Ibid., p. xxi.

114 **"cold, wet and polluting":** Ibid., p. xviii.

114 **"to leave a loved one's body:** Ibid.

114 **Whitehead argues that since the English:** Whitehead, "Carib cannibalism," pp. 78–80.

115 **"very difficult to track":** W. Raleigh, *The Discovery of the Large, Rich and Beautiful Empire of Guiana*, edited by R. Schomburgk (London: Hakluyt Society, 1868), p. 85.

115 **he was not part of the landing party:** P. Hulme, "Making sense of the native Caribbean," *New West Indian Guide (Nieuwe West-Indische Gids)* 67, nos. 3–4 (1993): 207.

116 **In 1828, author and historian Washington Irving:** W. Irving, "The life and voyages of Christopher Columbus," in *The Complete Works of Washington Irving*, vol. 11, edited by John Harmon McElroy (Boston: Twayne, 1981), p. 192. Originally published in 1828.

116 **In what would become a blueprint:** For example, F. MacNutt, *De Orbe Novo: The Eight Decades of Peter Martyr D'Anghera*, vol. 1 (New York: G. P. Putnam's Sons, 1912), p. 72; J. Blaine, J. Buel, J. Ridpath, and B. Butterworth, *Columbus and Columbia: A Pictorial History of the Man and the Nation* (Richmond, VA: B. F. Johnson, 1892), p. 172; J. Cummins, *Cannibals* (Guilford, CT: Lyons Press, 2001), p. x.; D. Korn, M. Radice, and C. Hawes, *Cannibal: The History of the People-Eaters* (London: Channel 4 Books, 2001), p. 11.

117 **It was then that Stony Brook:** William Arens, *The Man-Eating Myth* (Oxford: Oxford University Press, 1979), p. 206.

118 **"The most certain thing to be said:** Ibid., p.139.

118 **"unsophisticated":** I. Brady, Review of *The Man-Eating Myth*, *American Anthropologist* 84, no. 3 (1982): 595–611.

118 **"dangerous":** P. G. Riviere, Review of *The Man-Eating Myth*, *Man* 15, no. 1 (1980): 203–5.

118 **"does not advance our knowledge of cannibalism":** J. W. Springer, Review of *The Man-Eating Myth*, *Anthropological Quarterly* 53, no. 2 (1980): 148–50.

118 **Some critics also took the opportunity to attack Arens personally:** William Arens, with a reply by M. Sahlins, "Cannibalism: An exchange," *New York Review of Books* 26, no. 4 (March 22, 1979): 46–47.

118 **One anthropologist, who believed:** G. Obeyesekere, *Cannibal Talk: The Man-Eating Myth and Human Sacrifice in the South Seas* (Berkeley, CA: University of California Press, 2005), p. 2.

119 **"seldom the first outsiders to set foot:** Conklin, *Consuming Grief*, p. 14.

120 **Like the many priests:** Ibid., p. 6.

## Chapter 11—Cannibalism and the Bible

126 **"His body and blood are truly contained:** Legion of Mary (Tidewater, VA), "Fourth Lateran Council (1215)," http://www.legionofmarytidewater.com/faith/ECUM12.HTM.

126 **In 1520, though, Martin Luther:** M. Luther, *On the Babylonian Captivity of the Church* (October 6, 1520).

126 **In their entertaining book:** D. Diehl and M. Donnelly, *Eat Thy Neighbor: A History of Cannibalism* (Phoenix Mill, UK: Sutton, 2006), p. 22.

127 **In the celebration of [the Eucharist]:** J. H. Leith, ed., *Creeds of the Churches: A Reader in Christian Doctrine, from the Bible to the Present* (Louisville, KY: Westminster John Knox Press, 1982), pp. 502–3.

127 **Even as recently as 1965:** Pope Paul VI, "Mysterium Fedei: Encyclical of Pope Paul VI on the Holy Eucharist," September 3, 1965, http://www.vatican.va/holy_father/paul_vi/encyclicals/documents/hf_p-vi_enc_03091965_mysterium_en.html.

130 **In 1994, Dr. Johanna Cullen:** J. Cullen, "The miracle of Bolsena," *ASM News* 60 (1994): 187–91.

131 **The renowned organic chemist:** L. Garlaschelli, "Starch and hemoglobin: The miracle of Bolsena," *Chemistry and Industry* 80 (1998): 1201.

132 **"We were hungry, we were cold:** T. Taylor, *The Buried Soul* (Boston: Beacon Press, 2002), p. 74.

## Chapter 12—The Worst Party Ever

142 **Prolonged hunger carves the body:** S. A. Russell, *Hunger: An Unnatural History* (New York: Basic Books, 2005), p. 120.

144 **In 1980, anthropologist Robert Dirks wrote:** R. Dirks, "Social responses during severe food shortages and famine," *Current Anthropology* 21, no. 1 (1980): 21–44.

144 **In his book *The Cannibal Within*:** L. Petrinovich, *The Cannibal Within* (Piscataway, NJ: Transaction, 2000), p. 232.

144 **"It is not advantageous:** Ibid., p. 36.

145 **which came to be called the Minnesota Experiment:** L. Kalm and R. Semba, "They starved so that others be better fed: Remembering Ancel Keys and the Minnesota Experiment," *Journal of Nutrition* 135 (2005): 1347–52.

145 **"All offers of surrender from Leningrad:** M. Jones, *Leningrad: State of Siege* (New York: Basic Books, 2008); Hitler quote from inside cover.

145 **"Leningrad must die of starvation":** D. King, *Red Star Over Russia: A Visual History of the Soviet Union from the Revolution to the Death of Stalin* (New York: Harry N. Abrams, 2009), p. 318.

146 **According to historian David Glantz:** Ibid., p. 89.

146 **Archivist Nadezhda Cherepenina reported:** Nadezhda Cherepenina, "Assessing the scale of famine and death in the besieged city," in *Life and Death in Besieged Leningrad, 1941–44*, edited by John Barber and Andrei Dzeniskevich (Hampshire, UK: Palgrave MacMillan, 2005), p. 44.

146 **Pulitzer Prize–winning *New York Times* correspondent:** H. Salisbury, *The 900 Days: The Siege of Leningrad* (New York: Harper and Row, 1969), pp. 478–79.

146 **In a system designed to maximize:** M. Jones, *Leningrad*, pp. xxi–xxii.

146 **Rations were reduced a total of five times:** David M. Glantz, *The Siege of Leningrad* (London: Cassell Military Paperbacks, 2001), p. 83.

147 **"because their flesh was so much more tender":** Salisbury, *The 900 Days*, p. 479.

147 **"In the worst period of the siege":** Ibid., p. 478.

147 **According to numerous survivor accounts:** Jones, *Leningrad*, pp. 242–44.

148 **"[the men] noticed as they piled:** H. Askenasy, *Cannibalism: From Sacrifice to Survival* (New York: Prometheus Books, 1994), p. 75.

148 **"You will look in vain:** Salisbury, *The 900 Days*, p. 474.

148 **the official reports made right after the war:** D. Korn, M. Radice, and C. Hawes, *Cannibal: The History of the People-Eaters* (London: Channel 4 Books, 2001), pp. 83–86.

150 **Although the "poor Ethiopian" begged:** E. Leslie, "The ownership of the plank: David Harrison on the wreck of the *Peggy*," in *Cannibals: Shocking True Tales of the Last Taboo on Land and Sea*, edited by Joseph S. Cummins (Guilford, CT: Lyons Press, 2001), p. 50.

151 **Someone suggested that two of the men:** K. Johnson, "J. Quinn Thornton (1810-1888)," in *"Unfortunate Emigrants": Narratives of the Donner Party*, edited by Kristin Johnson (Logan, UT: Utah State University Press, 1996), p. 52.

152 **"his miserable companions cut the flesh:** Ibid., p. 53.

152 **"Being nearly out of provisions:** Edwin Bryant, *What I Saw in California* (1848; Lincoln: University of Nebraska Press, 1985), chap. 3.

152 **Foster, who survived the whole ordeal:** K. Johnson, "The Murphy family," New Light on the Donner Party (website), http://www.utahcrossroads.org/DonnerParty/Murphy.htm#William%20McFadden%20Foster.

154 **We looked all around but no living thing**: Esarey/Esrey/Esry Genealogy and Reunion (website), "Esarey-Esrey & Rhoads-Esrey letters: Records of a 19th century American migration," http://www.esarey.us/reunion/1873.htm.

156 **[Reed's] party immediately commenced distributing:** J. Merryman, "Adventures in California—Narrative of J. F. Reed—Life in the Wilderness—Sufferings of the Emigrants," *Illinois State Register*, November 19, 1847, p. 1, c. 6–7; p. 2, c. 1; November 26, 1847, p. 1, c. 2. Also published as "Narrative of the Sufferings of a Company of Emigrants in the Mountains of California, in the winter of '46 and '7 by J. F. Reed, late of Sangamon County, Illinois," *Illinois Journal* (Springfield), December 8, 1847, p. 1, c. 2–4.

156 **The mutilated body of a friend:** Johnson, "J. Quinn Thornton (1810–1888)," p. 90.

156 **They had consumed four bodies:** Ibid., p. 91.

157 **"No traces of her person could be found":** Dale Lowell Morton, ed., *Overland in 1846*, vol. 1: *Diaries and Letters of the California-Oregon Trail* (Lincoln: University of Nebraska Press, 1993), p. 362.

158 **"human skeletons . . . in every variety:** E. Rarick, *Desperate Passage* (New York: Oxford University Press, 2008), p. 229.

159 **"Analysis Finally Clears Donner Party:** R. Preidt, "Analysis finally clears Donner Party of rumored cannibalism," MedicineNet, April 19, 2010, http://www.medicinenet.com/script/main/art.asp?articlekey=115529.

160 **"Donners Ate Family Dog, Maybe Not People":** J. Viegas, "Donner Party ate family dog, maybe not people," Discovery News, April 15, 2010, http://news.discovery.com/history/donner-party-cannibalism.html.

160 **Even the *New York Times* got into the act:** S. Dubner, "No cannibalism among the Donner Party?" *Freakonomics* (*New York Times* blog), April 16, 2010, http://freakonomics.blogs.nytimes.com/2010/04/16/no-cannibalism-among-the-donner-party/.

160 **"Scientists Crash Donner Party":** *The Rat* (blog), "Scientists crash the Donner Party," January 13, 2006, http://therat.blogspot.com/2006_01_01_archive.html.

161 **Research conducted by Dr. Gwen Robbins:** Appalachian State University News, "Appalachian professor's research finds no evidence of cannibalism at Donner Party campsite," April 15, 2010, http://www.freerepublic.com/focus/chat/2493838/posts.

165 **"Professor's Research Demonstrates Starvation Diet:** Appalachian State University News, "Professor's research demonstrates starvation diet at the Donner Party's Alder Creek camp," April 23, 2010, http://www.news.appstate.edu/2010/04/23/professor%E2%80%99s-research-demonstrates-starvation-diet-at-the-donner-party%E2%80%99s-alder-creek-camp/.

165 **The actual research paper (published three months later:** K. J. Dixon, S. Novak, G. Robbins, J. Schablitsky, G. R. Scott, and G. Tasa, "'Men, women and children starving': Archaeology of the Donner family camp," *American Antiquity* 75, no. 3 (2010): 626–56.

166 **"'Cannibal' Doc Eats Her Words":** D. K. Li, "'Cannibal' doc eats her words," *New York Post*, April 22, 2010.

167 **In 1990, anthropologist Donald Grayson:** D. K. Grayson, "Donner Party deaths: A demographic assessment," *Journal of Anthropological Research* 46, no. 3 (1990): 223–42.

168 **When the Donner Party hacked a trail:** D. Grayson, "The timing of Donner Party deaths, Appendix 3," in *The Archaeology of the Donner*

*Party*, edited by Donald Hardesty (Reno: University of Nevada Press, 1997), p. 125.

169 **single and unpaired individuals are more responsive:** D. Maestripieri, N. Baran, P. Sapienza, and L. Zingales, "Between- and within-sex variation in hormonal responses to psychological stress in a large sample of college students," *Stress* 13, no. 5 (2010): 413–24.

169 **Grebenkemper told me that in 2011 and 2012:** J. Grebenkemper and K. Johnson, "Forensic canine search for the Donner family winter camps at Alder Creek," *Overland Journal* 33, no. 2 (Summer 2015): 65–89.

## Chapter 13—Eating People Is Bad

174 **Baby, baby, naughty baby:** Iona Opie and Peter Opie, eds., *The Oxford Dictionary of Nursery Rhymes* (Oxford: Oxford University Press, 1951), p. 59.

175 **"They immediately shewed [*sic*] as much horror:** G. Obeyesekere, "British cannibals: Contemplation of an event in the death and resurrection of James Cook, explorer," in *Beyond Textuality: Asceticism and Violence in Anthropological Interpretation*, edited by Gilles Bibeau and Ellen Corin (Berlin: Mouton de Gruyter, 1994), p. 146.

175 **"belief that a man needed his body after death:** R. Tannahill, *Flesh & Blood: A History of the Cannibal Complex* (New York: Little Brown, 1975), p. 45.

175 **"an unprecedented and almost pathological:** Ibid., p. 47.

176 **"you are what you eat":** M. Kilgore, "The function of cannibalism at the present time," in *Cannibalism and the Colonial World*, edited by F. Baker, P. Hulme, and M. Iversen (Cambridge, UK: Cambridge University Press, 1998), p. 239.

176 **"defined in terms of how:** Ibid., p. 239.

177 **Lurching up, he lunged out:** Homer, *The Odyssey*, translated by Robert Fagles (New York: Penguin Books, 1996), p. 222.

177 **"as wine came spurting:** Ibid., p. 223.

179 **"No price in the world":** Herodotus, *The History*, translated by David Grene (Chicago: University of Chicago Press, 1987), p. 228.

179 **"These are matters of settled custom":** Ibid.

179 **Herodotus was also the first writer:** C. Avramescu, *An Intellectual History of Cannibalism*, translated by Alister Ian Blyth (Princeton, NJ: Princeton University Press, 2009), p. 33.

179 **"some of them did something dreadful":** Herodotus, *The History*, p. 222.

179 **The Father of History also wrote extensively about the Scythians:** Ibid., pp. 303–4.

180 **"[made] water so greatly that she filled:** Ibid., p. 84.

180 **"his daughter's privy parts":** Ibid., p. 84.

181 **When Harpagus's son came to Astyages:** Ibid., pp. 89–90.

185 **as translated by Paul Larue:** J. Zipes, *Trials and Tribulations of Little Red Riding Hood* (New York: Routledge, 1993), p. 348.

185 **"cruel Ogre who eats little children":** Charles Perrault, ed., *The Tales of Mother Goose* (New York: D. C. Heath, 1901), p. 36.

186 **"great scarcity fell on the land":** Brothers Grimm, *Hansel and Gretel and Other Tales* (London: Sovereign, 2013), p. 7.

186 **"When he is fat I will eat him":** Ibid., p. 11.

186 **"Let Hansel be fat or lean:** Ibid., p. 11.

186 **According to Maria Tatar:** M. Tatar, *Off with Their Heads* (Princeton, NJ: Princeton University Press, 1992), p. 295.

187 **In Tabart's story, Jack is "indolent:** Maria Tatar, ed. and trans., *The Annotated Classic Fairy Tales* (New York: W. W. Norton and Company, 2002), p. 133.

187 **"My man is an ogre and there's nothing:** Ibid., p. 136.

187 **Fe-fi-fo-fum:** Ibid., p. 136.

188 **In Joseph Jacobs's revised epilogue, a "good fairy":** Ibid., 143.

188 **"Jack and his mother became very rich:** Ibid., p. 144.

188 **"the foundation stones of nursery literature:** M. Warner, "Fee fie fo fum: The child in the jaws of history," in Baker et al., *Cannibalism and the Colonial World*, p. 160.

189 **I was perfectly confounded and amazed:** D. Defoe, *Robinson Crusoe*, New York Post Family Classics Library (New York: Paperview Group, 2004), p. 122.

190 **"the wretched, inhuman custom of their devouring:** Ibid., p. 123.

190 **"justify the conduct of the Spaniards:** Ibid., p. 127.

191 **The place was covered with human bones:** Ibid., p. 153.

191 **Friday "still had a hankering stomach:** Ibid., p. 153.

191 **"would never eat man's flesh anymore":** Ibid., p. 157.

191 **"now a good Christian":** Ibid., p. 163.

191 **"Let fly . . . in the name of God":** Ibid., p. 173.

191 **"Defoe's work is an effective contribution:** F. Lestringant, *Cannibals: The Discovery and Representation of the Cannibal from Columbus to Jules Verne*, translated by Rosemarie Morris (Berkeley, CA: University of California Press, 1997), p. 141.

192 **"judging the savage by the standard:** J. G. Frazer, *The Golden Bough: A Study in Magic and Religion* (New York: MacMillan Publishing Company, 1922), p. 342.

192 **"the custom of tearing in pieces:** Ibid., p. 454.

193 **"The natives are superficially agreeable but they go in:** H. Lapsley, M. Mead, and R. Benedict, *The Kinship of Women* (Amherst, MA: University of Massachusetts Press, 2001), p. 217.

193 **"Cannibal savages as they were:** Avaramescu, *An Intellectual History of Cannibalism*, p. 160.

194 **"mankind's earliest festival":** S. Freud, *Totem and Taboo* (New York: W. W. Norton, 1950), pp. x, 142.

194 **As Stony Brook University anthropologist Bill Arens wrote:** W. Arens, *The Man-Eating Myth* (Oxford: Oxford University Press, 1979), p. 145.

## Chapter 14—Eating People Is Good

196 **"The Anonymous Conquistador":** Anonymous Conquistador, "The chronicle of the Anonymous Conquistador," in *The Conquistadores*, edited by Patricia de Fuentes (New York: Orion Press, 1963), pp. 165–81.

197 **"an institutionalized practice of consuming certain:** Key Ray Chong, *Cannibalism in China* (Wakefield, NH: Longwood Academic, 1990), p. 2.

197 **"publicly and culturally sanctioned":** Ibid., p. 2.

197 **Chong's investigation provided three examples of siege-related:** Ibid., pp. 45–46.

198 **153 and 177 incidents of war-related and natural disaster–related:** Ibid., pp. 160–61.

198 **in the 2008 book *Mubei* (*Tombstone*):** J. Yang, *Mubei* (*Tombstone*) (Hong Kong: Tiandi Chubanshe, 2008; reprinted 2010 [8th ed.]).

199 **"people ate tree bark, weeds, bird droppings:** Perry Link, "China: From famine to Oslo," *New York Review of Books* 58, no. 1 (January 13, 2011): 52.

199 **also believes that 36 million deaths:** R. MacFarquahar, "The worst man-made catastrophe, ever," *New York Review of Books* 58, no. 2 (February 10, 2011): 28.

200 **Mao decided to install an "improved" version:** J. Becker, *Hungry Ghosts: Mao's Secret Famine* (New York: Free Press, 1996), pp. 64–70.

201 **Another of Mao's brainstorms:** Ibid., 76–77.

201 **"Traveling around the region over thirty years later:** Ibid., pp. 118–19.

202 **"hate, love, loyalty, filial piety:** Chong, *Cannibalism in China*, p. 2.

202 **"Methods of Cooking Human Flesh":** Ibid., pp. 145–57.

202 **"Baking, Roasting, Broiling:** Ibid., p. 149.

202 **"children's meat was the best food of all:** Ibid., p. 137.

203 **In *Shui Hu Chuan* (*The Tales of Water Margins*):** Becker, *Hungry Ghosts*, p. 216.

203 **"five regional cuisines":** Chong, *Cannibalism in China*, p. 145.

203 **"many examples of steaming or boiling:** Ibid., p. 153.

203 **Prisoners of war were preferred ingredients:** Ibid., p. 153.

203 **Once victims had been subjected to criticism:** "The slaughter at Wuxuan, Zhengi Yi," from *Scarlet Memorial: Tales of Cannibalism in Modern China*, in *Cannibals: Shocking True Tales of the Last Taboo on Land and Sea*, edited by Joseph S. Cummins (Guilford, CT: Lyons Press, 2001), p. 210.

204 **"children would cut off parts of their body:** Chong, *Cannibalism in China*, p. 154.

204 **Far less frequent, but recorded nonetheless:** Ibid., pp. 100–101.

## Chapter 15—Chia Skulls and Mummy Powder

207 **Arens had previously acknowledged:** W. Arens, "Rethinking anthropophagy," in *Cannibalism and the Colonial World*, edited by Francis Barker, Peter Hulme, and Margaret Iverson (New York: Cambridge University Press, 1998), pp. 39–63.

207 **Referring to a reported instance of bone ash cannibalism:** G. Dole, "Endocannibalism among the Amahuaca Indians," *Transactions of the New York Academy of Science (Series II)* 24 (1962): 567–73.

209 **"although the theoretical possibility of customary:** W. Arens, *The Man-Eating Myth* (Oxford: Oxford University Press, 1979), p. 184.

210 **Scholar Key Ray Chong:** Chong, *Cannibalism in China* (Wakefield, NH: Longwood Academic, 1990), p. 93.

210 **"the gall bladder, bones, hair:** D. Korn, M. Radice, and C. Hawes, *Cannibal: The History of the People-Eaters* (London: Channel 4 Books, 2001), p. 92.

211 **So popular was this practice:** K. Gordon-Grube, "Anthropophagy in post-Renaissance Europe: The tradition of medicinal cannibalism," *American Anthropologist* 90 (1988): 407.

212 ***Mummies, Cannibals and Vampires*:** R. Sugg, *Mummies, Cannibals and Vampires: The History of Corpse Medicine from the Renaissance to the Victorians* (London: Routledge, 2011), p. 384.

212 **"human liver . . . oil distilled from human brains:** Richard Sugg, "The Aztecs, cannibalism and corpse medicine (1)," Aztecs at Mexicolore (website), http://www.mexicolore.co.uk/index.php?one=azt&two=aaa&id=325&typ=reg.

213 **"One thing we are rarely taught at school:** Fiona Macrae, "British royalty dined on human flesh (but don't worry, it was 300 years ago)," *Daily Mail*, May 21, 2011, http://www.dailymail.co.uk/news/article-1389142/British-royalty-dined-human-flesh-dont-worry-300-years-ago.html.

213 **Additional high-profile advocates of medicinal cannibalism included:** Sugg, "The Aztecs, cannibalism and corpse medicine (1)."

213 **With an ever-increasing demand:** Gordon-Grube, "Anthropophagy in post-Renaissance Europe," p. 407.

213 **"privy members cut off:** J. Lawrence, *A History of Capital Punishment* (New York: Citadel Press, 1960), p. 193.

213 **“upon some of the city gates”:** Gordon-Grube, “Anthropophagy in post-Renaissance Europe,” p. 407.

214 **Researcher Paolo Modenesi believes:** P. Modenesi, “Skull lichens: A curious chapter in the history of phytotherapy,” *Fitoterapia* 80 (2009): 145–48.

214 **Ideally, the moss from the skulls:** Gordon-Grube, “Anthropophagy in post-Renaissance Europe,” p. 406.

214 **“the cranium of a carcass that had been broken:** Modenesi, “Skull lichens,” p. 147.

215 **a bizarre medical treatment known as hoplochrisma:** Ibid., p. 147.

215 **“choose what is of a shining black:** A. Wootton, *Chronicles of Pharmacy*, 2 vols. (New York: USV Pharmaceutical Corporation, 1972), 2:24.

216 **[The Paracelist Oswald] Croll recommended:** Gordon-Grube, “Anthropophagy in post-Renaissance Europe,” p. 406.

217 **Listed as *mumia vera aegyptica*:** Modenesi, “Skull lichens,” p. 148.

217 **“the rise of Enlightenment attitudes:** Sugg, *Mummies, Cannibals and Vampires*, pp. 264–65.

217 **“after having Dad’s ashes:** K. Richards, *Life* (New York: Little Brown, 2010), p. 546.

## Chapter 16—Placenta Helper

219 **It gave me the wildest rush:** A. A. Abrahamian, “The placenta cookbook,” *New York*, August 29, 2011, p. 49.

219 **“Perfect,” “beautiful,” “precious”:** Ibid., p. 48.

224 **In 1930, primatologist Otto Tinklepaugh took a break:** O. Tinklepaugh and C. Hartman, “Behavioral aspects of parturition in the monkey (*Macacus rhesus*),” *Journal of Comparative Psychology* 11, no. 1 (1930): 63–98.

225 **Researchers initially posited:** M. B. Kristal, J. DiPirro, and A. Thompson, “Placentophagia in humans and nonhuman mammals: Causes and consequences,” *Ecology of Food and Nutrition* 51 (2012): 179.

225 **“voracious carnivorousness”:** Ibid., p. 179.

225 **He and his colleagues set out to investigate**: M. B. Kristal, A. Thompson, and H. Grishkat, “Placenta ingestion enhances opiate analgesia in rats,” *Physiology and Behavior* 35 (1985): 481–86.

226 **In 2010, researchers at the University of Nevada:** S. Young and D. Benyshek, "In search of human placentophagy: A cross-cultural survey of human placenta consumption, disposal practices, and cultural beliefs," *Ecology of Food and Nutrition* 49 (2010): 467–84.

227 **"throwing it into a lake":** Ibid., p. 473.

227 **followed by "burial":** Ibid., p. 473.

227 **"hanging or placing the placenta in a tree":** Ibid., p. 473.

227 **the *Great Pharmacopoiea of 1596*:** W. Ober, "Notes on placentophagy," *Bulletin of the New York Academy of Medicine* 55, no. 6 (1979): 596.

227 **"coldness of the sexual organs:** Ibid., p. 596.

228 **On a more recent and Western note:** S. Benet, *Song, Dance and Customs of Peasant Poland* (New York: Roy, 1951), pp. 196–97.

228 **"I Ate My Wife's Placenta:** N. Baines, "I ate my wife's placenta raw in a smoothie and cooked in a taco," *Guardian,* April 30, 2014, http://www.theguardian.com/lifeandstyle/2014/apr/30/i-ate-wifes-placenta-smoothie-taco-afterbirth.

234 **In a 1954 study, Czech researchers:** E. Soyková-Pachnerová, B. Golová, and E. Zvolská, "Placenta as a lactagogon," *Gynaecologia* 138 (1954): 617–27.

238 **The Somosomo people were fed:** Online Etymology Dictionary, "Extract of a letter from the Rev. John Watsford, dated October 6, 1846," in "Wesleyan Missionary Notices," September 1847, http://www.etymonline.com/index.php?term=long+pig&allowed_in_frame.

239 **"I sautéed the steak of Bernd with salt:** *Spiegel Online International,* "First TV interview with German cannibal: 'Human flesh tastes like pork,'" October 16, 2007, http://www.spiegel.de/international/zeitgeist/first-tv-interview-with-german-cannibal-human-flesh-tastes-like-pork-a-511775.html.

239 **compared his victim's flesh to raw tuna:** T. Kosuga, "Who's hungry?" translated by Lena Oishi, *Vice,* January 2, 2009, http://www.vice.com/read/whos-hungry-502-v16n1.

240 **"It was good, fully developed veal:** W. Seabrook, *Jungle Ways* (New York: Harcourt, Brace, 1931), p. 272.

241 **"Animals eat their placenta:** BBC News, "Why eat a placenta?" April 18, 2016, http://news.bbc.co.uk/2/hi/uk_news/magazine/4918290.stm.

## Chapter 17—Cannibalism in the Pacific Islands

244 **According to research biochemist Colm Kelleher:** C. Kelleher, *Brain Trust* (New York: Paraview Pocket Books, 2004), p. 115.

244 **By 1987, there were over 400 confirmed cases:** Ibid., p. 117.

247 **Fore elders told the foreigners:** W. Anderson, *The Collectors of Lost Souls* (Baltimore: Johns Hopkins University Press, 2008), p. 14.

248 **the locals were "difficult people to deal with":** Ibid., p. 21–22.

248 **"Actually these people are 'bestial' in many ways":** Ibid., p. 25.

248 **"Dead human flesh, to these people:** Ibid., p. 24.

249 **A decade later, the not-yet-controversial anthropologist:** W. Arens, *The Man-Eating Myth* (Oxford: Oxford University Press, 1979), p. 99.

249 **Arens was particularly galled by Berndt's description:** Ibid., p. 99.

249 **"Now you have cut off my penis!":** R. Berndt, *Excess and Restraint* (Chicago: University of Chicago Press, 1962), p. 283.

250 **He studied rabies and plague:** R. Rhodes, *Deadly Feasts* (New York: Simon and Schuster, 1997), p. 32.

250 **I am in one of the most remote:** Anderson, *The Collectors of Lost Souls*, p. 61.

251 **But Gajdusek had never seen any actual cannibalism:** Ibid., pp. 62–80.

251 **their initial findings were published in the prestigious:** D. C. Gajdusek and V. Zigas, "Degenerative disease of the central nervous system in New Guinea: The endemic occurrence of kuru in the native population," *New England Journal of Medicine* 257 (1957): 974–78.

252 **"The closest condition I can think:** Anderson, *The Collectors of Lost Souls*, p. 80.

252 **Another NIH scientist noticed a similarity:** Kelleher, *Brain Trust*, pp. 29–31.

252 **Scrapie, which was present in European sheep:** Ibid., pp. 31–32.

253 **Miffed that medical researchers were now intruding:** Anderson, *The Collectors of Lost Souls*, p. 81.

253 **Bennett proposed that a mutant kuru gene:** Ibid., p. 124.

254 **In the eastern highlands of New Guinea:** Anonymous, "The laughing death," *Time*, November 11, 1957, pp. 55–56, http://www.time.com/time/magazine/article/0,9171,867948-1,00.html#ixzz0sAV5Fs00.

254 **For his part, Gajdusek hated the media coverage:** Anderson, *The Collectors of Lost Souls*, p. 81.

256 **Finally, any society has practices considered:** J. Diamond, "Talk of cannibalism," *Nature* 407, no. 6800 (September 7, 2000): 25-26.

258 **Nevertheless, Robert Glasse published his and his then-wife's:** R. M. Glasse, "The social effects of kuru," *Papua New Guinea Medical Journal* 7 (1964): 36–47.

258 **Finally, Glasse calculated that kuru:** John D. Mathews, Robert Glasse, and Shirley Lindenbaum, "Kuru and cannibalism," *Lancet* 2 (1968): 449–52.

259 **Nearly 50 years after Glasse published:** J. Whitfield, W. Pako, J. Collinge, and M. Alpers, "Mortuary rites of the South Fore and kuru," *Philosophical Transactions of the Royal Society B* 363 (2008): 3721–24.

259 **As for how the funerary practices:** Ibid., p. 3722.

260 **"nothing was lost on the ground:** Ibid., p. 3722.

260 **The head of the deceased:** Ibid., p. 3724.

260 **including reproductive organs and feces:** Rhodes, *Deadly Feasts*, pp. 22–23.

261 **By tracing the path of the kuru reports:** S. Lindenbaum, "Cannibalism, kuru and anthropology," *Folia Neuropathologica* 47, no. 2 (2009): 139.

261 **Whitfield, who conducted nearly 200 interviews:** Shirley Lindenbaum and Jerome Whitfield, personal communications.

262 **In 2008, Michael Alpers wrote:** M. Alpers, "The epidemiology of kuru: Monitoring the epidemic from its peak to its end," *Philosophical Transactions of the Royal Society of London B, Biological Sciences* 363, no. 1510 (2008): 3707–13.

## Chapter 18—Mad Cows and Englishmen

263 **Unfortunately, the custom of consuming human flesh:** B. Fagan, *The Aztecs* (New York: W. H. Freeman, 1984), p. 233.

263 **In England, however, where there is no substantial soybean:** C. Kelleher, *Brain Trust* (New York: Paraview Pocket Books, 2004), p. 118.

264 **In searching for answers, the British government enlisted:** P. Yam, *The Pathological Protein* (New York: Copernicus Books, 2003), p. 110.

265 **Previously, dangerous solvents had been used:** Kelleher, *Brain Trust*, p. 120.

265 **The first was a significant increase:** Rhodes, *Deadly Feasts* (New York: Simon and Schuster, 1997), p. 180.

266 **In 1947, an outbreak of what would become known**: Rhodes, *Deadly Feasts*, pp. 81–82.

267 **they inoculated a trio of chimpanzees:** Ibid., p. 87.

269 **In a February 1966 article:** D. C. Gajdusek, C. Gibbs, and M. Alpers. "Experimental transmission of a kuru-like syndrome to chimpanzees," *Nature* 209 (February 19, 1966): 794–96.

269 **"The mechanism of spread of kuru:** D. C. Gajdusek, "Kuru in the New Guinea highlands," in *Tropical Neurology*, edited by J. D. Spillane (New York: Oxford University Press), pp. 376–83.

270 **In a 2002 interview, the NIH researcher:** R. Uridge, "BSE: The untold story," Fortune City, January 23, 2002, http://www.fortunecity.com/emachines/e11/86/bse.html.

271 **At the forefront of the mystery:** Rhodes, *Deadly Feasts*, pp. 120–22.

272 **In another set of experiments, South African radiation biologist:** T. Alper, D. Haig, and M. Clarke, "The exceptionally small size of the scrapie agent," *Biochemical and Biophysical Research Communications* 22 (1966): 278–84.

272 **After reading over Alper's work:** J. S. Griffith, "Self-replication and scrapie," *Nature* 215, no. 5105 (1967): 1043–44.

272 **But Stanley Prusiner, a young biochemist:** Anderson, *The Collectors of Lost Souls*, pp. 190–94.

273 **In 1982, Prusiner published his lab findings:** S. Prusiner, "Novel proteinaceous infectious particles cause scrapie," *Science* 216, no. 4542 (1982): 136–44.

274 **"invaded and colonized the work:** Rhodes, *Deadly Feasts*, p. 203.

274 **accusations that he had used the peer review:** Ibid., pp. 203–5.

## Chapter 19—Acceptable Risk

276 **a "blue ribbon" panel:** C. Kelleher, *Brain Trust* (New York: Paraview Pocket Books, 2004), p. 125.

277 **offering them only 50 percent of market value:** P. Yam, *The Pathological Protein* (New York: Copernicus Books, 2003), p. 118.

277 **"Spongiform Fear Grows":** L. Cahill, "Spongiform fear grows," *Farming News*, April 22, 1988.

277 **"Raging Cattle Attacks":** D. Brown, "Raging cattle attacks," *Sunday Telegraph*, April 24, 1988.

277 **An earlier paper in *Nature* also demonstrated:** C. Gibbs Jr. and D. C. Gajdusek, "Transmission of scrapie to the cynomolgus monkey (*Macaca fascicularis*)," *Nature* 236 (1972): 73–74.

278 **"the risk of transmission of BSE:** Southwood Working Party, *The BSE Inquiry*, vol. 4: *The Southwood Working Party, 1988–89*, National Archives, http://collections.europarchive.org/tna/20090505194948/http://bseinquiry.gov.uk/report/volume4/chapt102.htm#888456.

278 **The authors of the Southwood report:** Kelleher, *Brain Trust*, p. 140.

278 **In 2007, she and her coworkers identified:** L. Manuelides, Y. Zhoa-Xue, N. Barquero, and B. Mullins, "Cells infected with scrapie and Creutzfeldt-Jakob disease agents produce intracellular 25-nm virus-like particles," *Proceedings of the National Academy of Sciences* 104, no. 6 (2007): 1966–70.

280 **In 1993, two British dairy farmers died:** Rhodes, *Deadly Feasts* (New York: Simon and Schuster, 1997), p. 187.

280 **In May of the same year, 15-year-old Victoria Rimmer:** Ibid., pp. 187–88.

281 **"Think about the economy":** Ibid., p. 188.

281 **In 1994, a 16-year-old schoolgirl:** Ibid., p. 188–89.

281 **On March 8, 1996, the hammer fell:** Kelleher, *Brain Trust*, p. 165.

281 **The new disease was initially:** J. Ironside and J. Bell, "Florid plaques and new variant Creutzfeldt-Jakob disease," *The Lancet* 350, no. 9089 (November 15, 1997): 1475.

281 **By October 2013, the number of definite:** A. Hodgekiss and J. Hope, "One in 2,000 people in the UK carry 'abnormal proteins' linked to mad cow disease," *Daily Mail*, October 15, 2013, http://www.dailymail.co.uk/health/article-2461354/Mad-cow-disease-One-2-000-people-UK-carry-abnormal-proteins-linked-vCJD.html

282 **"anonymous appendix samples:** *British Medical Journal*, "Researchers estimate one in 2,000 people in the UK carry variant CJD proteins," October 14, 2013, http://www.bmj.com/press-releases/2013/10/14/researchers-estimate-one-2000-people-uk-carry-variant-cjd-proteins.

282 **The researchers and their colleagues:** S. Mead, J. Whitfield, M. Poulter, P. Shah, J. Uphill, J. Beck, T. Campbell, H. Al-Dujaily, M. Alpers, and

J. Collinge, "Genetic susceptibility, evolution and the kuru epidemic," *Philosophical Transactions of the Royal Society, B* 363 (2008): 3741–46.

282 **In 2015, Collinge and his research team:** E. Asante, M. Smidak, A. Grimshaw, R. Houghton, A. Tomlinson, A. Jeelani, T. Jakubcova, S. Hamdan, A. Richard-Londt, J. Linehan, S. Brandner, M. Alpers, J. Whitfield, S. Mead, J. Wadsworth, and J. Collinge, "A naturally occurring variant of the human prion protein completely prevents prion disease," *Nature* 522, no. 7557 (2015): 478–81.

285 **"I raise the virus issue":** Rhodes, *Deadly Feasts*, p. 251.

# Epilogue—One Step Beyond

289 **"an older homosexual tramp:** J. Green, *Cassell's Dictionary of Slang* (London: The Orion Publishing Group, 2005), p. 240.

289 **"He is guilty of nothing more:** S. Larimer, "New York's 'cannibal cop' released after judge overturns 2013 conviction," *Washington Post*, July 1, 2014, http://www.washingtonpost.com/news/post-nation/wp/2014/07/01/new-yorks-cannibal-cop-released-after-judge-overturns-2013-conviction/.

290 **"are people who have a tremendous desire to destroy:** B. Robinson, "Why killers cannibalize," ABC News, May 22, 2002, http://abcnews.go.com/US/story?id=90012.

291 **"Viewing anything that involves violence:** F. Rice, "Whodunnit then? Here's why we're all so obsessed with violent crime," *Marie Claire*, March 9, 2015, http://www.marieclaire.co.uk/blogs/548712/whodunnit-then-here-s-why-we-re-all-so-obsessed-with-violent-crime.html.

291 **In the Darfur region of the Sudan:** O. Brown and R. McLeman, "Climate change as the 'new' security threat: Implications for Africa," *International Affairs* 83, no. 6 (2007): 1141–54.

291 **where researchers Daniel Griffin and Kevin Anchukaitis used soil:** D. Griffin and K. Anchukaitis, "How unusual is the 2012-2014 California drought?" *Geophysical Research Letters* 41 (2014): 9017–23.

292 **According to the United Nations Environmental Programme:** Ibid., p. 1143.

294 **as well as in "ancient Egypt, ancient Greece and Rome:** L. Petrinovich, *The Cannibal Within* (Piscataway, NJ: Transaction, 2000), p. 171.

# Recommended Books on Cannibalism and Related Topics

Warwick Anderson, *The Collectors of Lost Souls* (Baltimore: Johns Hopkins University Press, 2008).

William Arens, *The Man-Eating Myth* (Oxford: Oxford University Press, 1979).

Hans Askenasy, *Cannibalism—From Sacrifice to Survival* (New York: Prometheus Books, 1994).

Catalin Avramescu, *An Intellectual History of Cannibalism*, translated by Alister Ian Blyth (Princeton, NJ: Princeton University Press, 2009).

John Barber and Andrei Dzeniskevich, eds., *Life and Death in Besieged Leningrad, 1941–44* (Hampshire, UK: Palgrave MacMillan, 2005).

Francis Barker, Peter Hulme, and Margaret Iversen, eds., *Cannibalism and the Colonial World* (Cambridge, UK: Cambridge University Press, 1998).

Jasper Becker, *Hungry Ghosts: Mao's Secret Famine* (New York: The Free Press, 1996).

Ronald M. Berndt, *Excess and Restraint* (Chicago: University of Chicago Press, 1962).

James Gillespie Blaine, James William Buel, John Clark Ridpath, and Benjamin Butterworth, *Columbus and Columbia: A Pictorial History of the Man and the Nation* (Richmond, VA: B.F. Johnson and Company, 1892).

Robert Bloch, *Psycho* (New York: Random House, 1958).

Daniel James Brown, *The Indifferent Stars Above* (New York: William Morrow, 2009).

Robert Chambers, *Vestiges of the Natural History of Creation* (London: John Churchill, 1844).

Key Ray Chong, *Cannibalism in China* (Wakefield NH: Longwood Academic, 1990).

Beth A. Conklin, *Consuming Grief: Compassionate Cannibalism in an Amazonian Society* (Austin, TX: University of Texas Press, 2001).

Nathan Constantine, *A History of Cannibalism: From Ancient Cultures to Survival Stories and Modern Psychopaths* (Edison, NJ: Cartwell Books, 2006).

Homer Croy, *Wheels West* (New York: Hastings House, 1955).

Joseph S. Cummins, ed., *Cannibals: Shocking True Tales of the Last Taboo on Land and Sea* (Guilford, CT: The Lyons Press, 2001).

Francis Darwin, ed., *The Life and Letters of Charles Darwin, Including an Autobiographical Chapter* (London: John Murray, 1887).

Daniel Defoe, *Robinson Crusoe*, New York Post Family Classics Library (New York: Paperview Group, 2004).

Patricia de Fuentes, ed., *The Conquistadores* (New York: Orion Press, 1963).

Daniel Diehl and Mark P. Donnelly, *Eat Thy Neighbor—A History of Cannibalism* (UK: Sutton, 2006).

Niles Eldridge and Ian Tattersall, *The Myths of Human Evolution* (New York: Columbia University Press, 1982).

Mark Elgar and Bernard Crespi, eds., *Cannibalism: Ecology and Evolution among Diverse Taxa* (New York: Oxford University Press, 1992).

Brian M. Fagan, *The Aztecs* (New York: W.H. Freeman and Company, 1984).

James George Frazer, *The Golden Bough: A Study in Magic and Religion* (New York: MacMillan Publishing Company, 1922).

David M. Glantz, *The Siege of Leningrad* (London: Cassell Military Paperbacks, 2001).

Laurence R. Goldman, ed., *The Anthropology of Cannibalism* (Connecticut: Bergin and Garvey, 1999).

Leon Goure, *The Siege of Leningrad* (California: Stanford University Press, 1962).

Donald Hardesty, *Archaeology of the Donner Party* (Reno: Univ. of Nevada Press, 1997).

Lansford W. Hastings, *The Emigrants Guide to Oregon and California* (Bedford, MA: Applewood Books, 1994).

Glenn Hausfeter and Sarah Blaffer Hardy, eds., *Infanticide: Comparative and Evolutionary Perspectives* (Hawthorne, NY: Aldine Publishing, 1984).

Herodotus, *The History*, Translated by David Grene (Chicago: University of Chicago Press, 1987).

Homer, *The Odyssey*, translated by Robert Fagles (New York: Penguin Books, 1996).

Peter Hulme and Neil Whitehead, eds. *Wild Majesty: Encounters with Caribs from Columbus to the Present Day* (Oxford: Clarendon Press, 1992).

Yang Jisheng, *Mubei (Tombstone)* (Hong Kong: Tiandi Chubanshe, 2008), 1095 pp.; eighth edition (2010).

Kristin Johnson, ed., *"Unfortunate Emigrants": Narratives of the Donner Party* (Logan, UT: Utah State University Press, 1996).

Michael Jones, *Leningrad: State of Siege* (New York: Basic Books, 2008).

Colm A. Kelleher, *Brain Trust* (New York: Paraview Pocket Books, 2004).

Joseph King, *Winter of Entrapment*, (Toronto: P.D. Meany Publishers, 1992).

Robert Klitzman, *The Trembling Mountain* (New York: Plenum Trade, 1998).

Daniel Korn, Mark Radice, and Charlie Hawes, *Cannibal—The History of the People-Eaters* (London: Channel 4 Books, 2001).

Janet Leonard and Alex Córdoba-Aguilar, ed., *The Evolution of Primary Sexual Characteristics in Animals* (New York: Oxford University Press, 2010).

Frank Lestringant, *Cannibals: The Discovery and Representation of the Cannibal from Columbus to Jules Verne* (Berkeley, CA: University of California Press, 1997.

Francis Augustus MacNutt, *De Orbe Novo; The Eight Decades of Peter Martyr D'Anghera*, vol. 1 (New York and London: G.P. Putnam's Sons, 1912).

John Harmon McElroy, ed., *The Complete Works of Washington Irving*, vol. XI, (Boston: Twayne, 1981).

C.F. McGlashan, *History of the Donner Party: A Tragedy of the Sierra* (Stanford, CA: Stanford University Press, 1947).

Moira Martingale, *Cannibal Killers: The History of Impossible Murders* (New York: Carroll & Graf Publishers, 1993).

J.H.P. Murray, *Papua or British New Guinea* (London: T. Fisher Unwin, 1912).

Linda A. Newson, *Aboriginal and Spanish Colonial Trinidad*, (London: Academic Press, 1976).

Ganeth Obeyesekere, *Cannibal Talk: The Man-Eating Myth and Human Sacrifice in the South Seas* (Berkeley, CA: University of California Press, 2005).

Lewis Petrinovich, *The Cannibal Within* (Piscataway, NJ: Transaction, 2000).

Sheldon Rampton and John Stauber, *Mad Cow U.S.A.* (Monroe, MA: Common Courage Press, 2004).

Richard Rhodes, *Deadly Feasts* (New York: Simon and Schuster, 1997).

Irving Rouse, *The Tainos: Rise and Decline of the People Who Greeted Columbus* (Hartford and London: Yale University Press, 1992).

Sharman Apt Russell, *Hunger: An Unnatural History* (New York: Basic Books, 2005).

Harrison Salisbury, *The 900 Days: The Siege of Leningrad* (New York: Harper and Row, 1969).

Carl O. Sauer, *The Early Spanish Main* (California: Univ. of California Press, 1966).

Bill Schutt, *Dark Banquet: Blood and the Curious Lives of Blood-Feeding Creatures* (New York: Harmony Books, 2008).

Ralph S. Solecki, *Shanidar: The First Flower People* (New York: Alfred A. Knopf, 1971).

David E. Stannard, *American Holocaust* (New York: Oxford University Press, 1992).

George R. Stewart, *Ordeal by Hunger* (New York: Houghton Mifflin, 1992).

Reay Tannahill. *Flesh & Blood—A History of the Cannibal Complex* (New York: Little Brown, 1975).

Maria Tatar, *Off With Their Heads* (Princeton, NJ: Princeton University Press, 1992).

Ian Tattersall, *The Fossil Trail* (New York: Oxford University Press, 2009).

Carole A. Travis-Henikoff, *Dinner with a Cannibal* (Santa Monica, CA: Santa Monica Press, 2008)

Philip Yam, The Pathological Protein (New York: Copernicus Books, 2003).

Zheng Yi, *Scarlet Memorial: Tales of Cannibalism in Modern China* (Boulder, CO: Westview Press, 1996).

Jack Zipes, *The Brothers Grimm: From Enchanted Forests to the Modern World* (London: Routledge Kegan & Paul, 1988).